The Origin of Everything

Uniting Science and Philosophy

by: *David Rowland*

Revised Edition: 2019

Copyright © DW Rowland Enterprises Inc. 2017, 2019

All rights reserved. No part of this book may be used or reproduced by any means, graphic, electronic, or mechanical, including photocopying, recording, taping, or by any information storage retrieval system without the written permission of the author – except in the case of brief quotations embodied in critical articles and reviews.

DW Rowland Enterprises Inc.
P.O. Box 30015, Prospect Plaza
Fredericton, NB E3B 3C2
david222@hush.com

ISBN: 9 7819 8088 9533

Contents

A Theory of Everything 5
Science and Philosophy Entwined 11
Scientific Speculation 31
Bang Goes the Theory 41
A Tale of Two Theories 59
Expanding or Infinite 63
Continuous Creation 67
Time is a Thought 75
Endless Space 81
Looking Back in Time 85
Faster than Light 89
A Universe of Attraction 103
A Universe of Light 107
Quantum Theory Demystified 109
Life Emerges 125
Evolutionary Anomalies 139
A Universe of Intention 153
An Infinite Universe 159
Consciousness 161
Consciousness Creates 167
Interesting Parallels 169
Glossary 171
About the Author 185

A Theory of Everything

Science tells us what happens.
Philosophy tells us why.

Stephen Hawking once said, "I am hopeful that we will find a consistent model that describes everything in the universe." His hope was to be able to develop an all-encompassing theory of everything – a coherent theoretical framework of physics that fully explains and mathematically links together all aspects of the universe and can be expressed in very few equations. Physics, however, is not everything.

Physics describes how a big bang supposedly created the elements and the cosmos. Chemistry describes how elements combined to form organic substances. Biology describes the behavior of living things whose structures are formed from organic molecules. Evolution describes how primitive living things developed into complex life forms, some of whom are conscious. The assumption is that physics set everything into motion. If so, then one would expect a theory of everything to be explicable entirely in terms of physics. It cannot be done, however. "Everything" is much bigger than any science or all combined sciences can explain.

The belief that physics set everything into motion is misleading. Physics is simply the mathematical language we use to describe physical phenomena that have already been created and how they interact. Physics is an effect, not a cause. Until we know what set physics into motion, we have no clue as to how the universe was created.

Science cannot explain why (or even if) there may have been a big bang, why there is physics, why there is gravity, how inert chemicals were transformed

there is gravity because space moves expands contracts curves into living beings, or how some of these beings acquired consciousness. Only philosophy can answer such questions.

Philosophy is the study of the fundamental nature of knowledge, reality and existence. One could say that philosophy is the mother of all studies. Philosophy gave us ethics, which gave us law. Philosophy gave us logic, which gave us mathematics. Until the 19th century, science was called "natural philosophy". Until the 1870s, psychology was a branch of philosophy. Perhaps the Doctor of Philosophy (PhD) degree, the highest one can earn in every academic field, is an acknowledgement that philosophy came first and encompasses everything.

The term "spontaneously" is often used to gloss over significant gaps in scientific understanding. Examples: the universe spontaneously created itself from nothing – inanimate matter spontaneously became living organisms – higher life forms spontaneously developed consciousness. None of these events is possible, however. Nothing cannot be the cause of something. Chemistry cannot create biology. Biology cannot create consciousness. Something beyond science is responsible for all the above transformations.

Everything which exists is real, but not everything that is real is physical. Science is limited to studying the structure and behavior of physical reality. Whatever is real that science cannot explain, therefore, must be nonphysical.

The physical universe is the consequence of a reality beyond itself – something nonphysical. Science can measure the effects that nonphysical reality has on our physical world. However, nonphysical reality itself cannot be measured. It simply is.

There is a nonphysical form of primary energy that both created the universe and continues to interact with it in subtle ways. Understanding that this is so unravels cosmological mysteries, resolves the quantum enigma, explains how life evolved from inanimate matter, expands our understanding of DNA, fills in gaps in evolutionary theory, explains how humans acquired consciousness, and predicts future scientific discoveries.

To develop a plausible theory of everything requires unifying science and philosophy. Science can tell us what happens in our physical world. Philosophy can tell us why. Unfortunately, these two disciplines which began together have grown apart. To re-unite them may be challenging to old belief patterns.

Cosmology

Cosmology is the study of the origin and development of the universe. Physical cosmology describes the physical origins and development of the universe on a large scale. Metaphysical cosmology addresses questions that are beyond the scope of science, such as "what is the reason for existence", "what is the ultimate first cause of everything", and "does consciousness have a purpose". The ancient Greeks made no distinction between physical and metaphysical cosmology, considering them to be one and the same.

Theories in physical cosmology depend on assumptions that cannot be tested – the primary one being that the observable universe is the entire universe. There may be much more going on than our scientific instrumentation can detect.

Physics and Metaphysics

Physics is the science dealing with the properties and interactions of matter and energy. Metaphysics may be considered the philosophy of physics. [*"Meta" is a word of Greek origin implying "beyond" or "of a higher order".*]

A plausible theory of everything must include and describe all aspects of reality. Physics is limited to that which can be measured. Whatever cannot be measured thus falls to metaphysics, the branch of philosophy which deals with the first principles of things (e.g., being, knowing, substance, essence, cause, time, and space).

A theory of everything thus requires the integration of physics with metaphysics. The language of physics is mathematics, the deductive science of number, quantity, space, and arrangement. The language of metaphysics is logic, the science of reasoning and thinking.

Physics answers questions that metaphysics cannot, and vice versa. The harmonious blending of these two disciplines requires that the knowledge gained from one complements or adds to the other, without conflict.

In their book, *Quantum Enigma*, Bruce Rosenblum and Fred Kuttner state: *"Something beyond ordinary physics awaits discovery … the existence of an enigma isn't a physics question, it's metaphysics in the original sense of the word … When it comes to metaphysics, non-physicists with a general understanding of the experimental facts – facts about which there is no dispute – can have an opinion matching that of physicists."*

Enigmas

If we look only to physics and mathematics to explain the universe, we encounter enigmas and paradoxes that science cannot explain. Examples of these include:

- If the universe had a beginning, when could this have happened? Time is in the universe; the universe is not in time. There is no point in time at which time began.

- If the universe had a beginning, from what was it made? There is no form of existence which predates existence.

- If the universe began with a "big bang" in which a single point (or microdot) violently exploded out pure energy which almost instantly became mass bearing particles that eventually formed stars, planets, and galaxies – how is such an origin possible? A point is an abstraction that refers to a specific location on a graph. Points do not in fact exist, having zero size and zero mass. How can nothing be the cause of everything?

- If the universe began from a single point, where in space could this point have been located? Space refers to the relative position of objects. Without matter, neither space nor location has any meaning.

- If electrons spontaneously disappear from one orbital and a tiny increment of time later show up in another orbital, where are these electrons when they are not physically present?

- If all species come into being by the evolution of earlier forms of life over successive generations, and if all species can ultimately be traced back to a single ancestor, where did that original life form (who was without ancestor) come from? However this came to be, could it have happened multiple times?

Metaphysics can explain the above mysteries. Surprisingly (or not), the same answer is at the root of every explanation: there is a nonphysical reality that interfaces with and directs the development of the physical universe and everything in it.

Philosophy gave us logic, which gave us mathematics. Philosophers have abandoned science, either because they do not understand the math or because they do not see its relevance. Science has retreated from philosophy into a world of theoretical mathematical models that often bear little relation to reality. Mainstream scientists are more concerned with making their equations work out than they are with the meaning that can be attached to them. It is as if math is being used to discredit logic. It is time to bring these two disciplines back together again.

 The only viable theory of everything is one that explains how a nonphysical source of energy manifests into and affects every aspect of our physical reality. Read on.

Science and Philosophy Entwined

Science and philosophy are complementary disciplines both seeking the same ultimate truths.

Who are we? Where did we come from? What is the nature of reality? Why does the universe exist? What is our part in the cosmos? For centuries humankind has been struggling to find the answers to these and similar philosophical questions.

Philosophy is the use of reason and argument in seeking truth and knowledge of reality, especially of the causes and nature of things and of the principles governing existence, the material universe, and the perception of physical phenomena. Eighteenth century philosophers considered all human knowledge, including science (which was then called "natural philosophy"), to be their field. Since the nineteenth century, however, science became specialized and divided nature into compartments that are often disconnected. Science also relies heavily on mathematics, but math cannot answer the question of why there is a universe.

Philosophy gave us logic – the science of reasoning, proof, thinking, or inference. Mathematics is a specialized form of logic – the deductive science of number, quantity, space, and arrangement. A plausible theory of everything needs to be understood by everyone, not just scientists. Logic is the common language that makes this possible. Logic is capable of answering the deeper questions that mathematics cannot.

Science studies "how" things happen. Philosophy explains "why" they happen. As such, philosophy is the science of the possible.

Although modern scientists come from one perspective and philosophers from another, there are some common themes among their various theories – such as that there may be (a) a nonphysical realm that transcends time and space, (b) a universal connectedness among all things, (c) a strong mind-world connection, (d) a fundamental consciousness, and (e) infinite possibilities.

Some philosopher/scientists believe that there is a nonphysical force which created our physical world [*Plato, Anaximander, Wheeler*]. Quantum theorists see evidence of this nonphysical energy at work at the subatomic level, guiding and connecting particles in ways that appear random, forming a world of possibilities rather than one of facts [*Heisenberg, Bell, Bohm, Clauser*]. Some believe that we live in a mental world that functions according to a deliberate design, and that the universe looks more like a great thought than a great machine, [*Aristotle, Berkeley, Jeans, Dyson*]. Many conclude that we live in a consciousness based reality, that consciousness is the fundamental principle of existence and the source of all being, and that consciousness and its contents are all that exist [*Schroedinger, James, Planck, Everett, Stapp, Goswami, Hoffman, Kafatos, Lind, Chalmers*].

Pythagoras (*570-495 BCE*), a mathematician and scientist, was the first person to call himself a philosopher, or lover of wisdom. Pythagoras was also the first to refer to the universe as a cosmos - meaning a complex and orderly, harmonious system.

Plato (*427-347 BCE*) was a philosopher and mathematician who believed in abstract realism, the idea that ***there are nonmaterial "objects" which exist in a third realm that transcends time and space***. He distinguished between the reality which is perceptible but unintelligible – and the reality which is imperceptible but intelligible. Plato believed that mathematics exists in an abstract space, that mathematical patterns predate the universe – suggesting that the universe was created by a universal mind.

Anaximander (*611-546 BCE*) was a philosopher and proponent of astronomy, physics, and geometry. He may have conducted the earliest recorded scientific experiment. Anaximander believed that the universe is infinite in extension and duration. He said, "***What is infinite is something other than the elements, and from it the elements arose.***" Anaximander also observed that if the first human had somehow appeared on earth as an infant, it would not have survived because humans are helpless at birth.

Parmenides (*515-450 BCE*) was a philosopher who believed that reality (which he called "what is") is one being -- that existence is timeless – and that the cosmos is uncreated, eternal, and indestructible. His statement, "For it is the same thing that can be thought and that can be", suggests that thoughts create.

Aristotle (*384-322 BCE*) was a philosopher whose writings include logic, metaphysics, ethics, aesthetics, poetry, theatre, music, rhetoric, linguistics, politics and government. Aristotle's works contain the earliest known formal study of logic, which was incorporated in the late 19th century into modern formal logic. In metaphysics, Aristotle influenced Judeo-Islamic philosophical and theological thought during the Middle Ages and continues to influence Christian theology today.

Aristotle was also the first genuine scientist in history. He wrote about physics, cosmology, biology, and zoology. Aristotle's lectures followed a pattern that formed the basis of the scientific method. He decided that the principles of nature could be found within nature and could be discovered using careful observation and inductive reasoning. Aristotle believed in ***"an intellectual natural world that functions according to some deliberate design."***

In his work, "Physics", Aristotle examined the nature of matter, space, time, and motion. His remarkable contributions to physics laid the groundwork for Galileo, Newton and Einstein. Aristotle gave us a model of the universe whose laws are invariant and capable of being discovered by observation and understood by reason – a model which lasted almost 20 centuries without significant modification. **Aristotle believed that the universe always existed, so he did not concern himself with the issue of how it came to be.**

St. Augustine (*354-430*) was a Christian philosopher who answered the question, "If the world was created a finite time ago, what happened before that?" Augustine's answer was that time was a property of the world that God created and that time did not exist before the creation. What he was really saying is that *time is in the universe, the universe is not in time*.

René Descartes (*1596-1650*) was a philosopher, mathematician, and scientist. He formulated the first modern version of mind-body metaphysical dualism – in which thinking is the essence of mind, and extension into three dimensions is the essence of matter.

Isaac Newton (*1642-1726*) was described in his own day as a "natural philosopher". He was a devout but unorthodox Christian who dedicated much of his time to the study of biblical chronology and alchemy. **Newton believed that our solar system did not arise out of chaos but was created by God, and that God can and does intervene in the workings of the universe.**

Best known as a physicist and mathematician, Newton laid the foundation for classical mechanics, made contributions to optics, and shares credit for the development of calculus. He believed that *"Truth is ever to be found in simplicity and not in the multiplicity and confusion of things."*

Newton formulated the laws of motion and universal gravitation, which dominated scientists' view of the physical universe for the next three centuries. He built the first practical reflecting telescope, developed a theory of color based on prismatic refraction of light, formulated an empirical law of cooling, studied the speed of sound, and introduced the concept of fluid viscosity.

Newton invented the mathematics that showed numerically how objects react when the force of gravity pulls on them. He claimed that the laws he discovered applied to everything in the universe. This was the beginning of both modern physics and modern astronomy.

Bishop George Berkeley (*1685-1723*) was a philosopher who advanced a theory he called "immaterialism" (later referred to as "subjective idealism" by others). This theory contends that physical objects are only ideas in the minds of perceivers and cannot exist without being perceived. As such, all material objects and all space and time are an illusion. Berkeley's bottom line was that ***nothing exists except the mind and its ideas***.

Berkley discussed the limitations of human vison and advanced the theory that the proper objects of sight are not material objects but light and color. He argued against Isaac Newton's doctrine of absolute space, time and motion – and in only that very limited sense was a precursor to the views of Einstein.

Christian Wolff (*1679-1754*) was an eminent philosopher who wrote on almost every scholarly subject of his time. He redefined philosophy as the "science of the possible". Wolff also wrote the first paper on cosmology, entitled *Cosmologia Generalis*, published in 1731.

Immanuel Kant (*1724-1804*) proposed the nebular hypothesis – that far in the past the universe was a nearly homogeneous, infinite gas that coalesced into vast agglomerations of stars what we now call galaxies. Kant believed that creation was and is a continuous process which spreads throughout the universe.

Charles Darwin (_1809-1882_) was a naturalist and geologist best known for his contribution to evolutionary theory. He believed that all species of life have descended over time from common ancestors, by the process of natural selection: "_I believe that animals have descended from at most only four or five progenitors and plants from an equal or lesser number._"

Although Darwin denied that God had a direct hand in creating species, he nonetheless indicated that God created the natural laws of the cosmos, including the laws of evolutionary development. He also made clear that **natural selection does not address the question of how species acquired mental abilities**: "_I must premise that I have nothing to do with the origin of primary mental powers, any more that I have with that of life itself. We are concerned only with the diversities of instinct and of other mental qualities of animals within the same class._"

William James (_1842-1910_) was a philosopher, psychologist and physician. **He believed in a mind-world connection which he described as a "stream of consciousness**", or flow of thoughts in the conscious mind.

Nikola Tesla (_1856-1943_) was a physicist, inventor, electrical engineer, mechanical engineer and futurist best known for his contribution to the design of the modern alternating current electricity supply system. Tesla said, "_If you want to find the secrets of the universe, think in terms of energy, frequency, and vibration,_" and, "**To me the universe is simply a great machine which never came into existence and never will end**."

Arthur Eddington (_1882-1944_) was an astronomer, physicist and mathematician who believed that physical reality is a product of mind. He said, "**The universe is of the nature of a thought or sensation in a universal Mind** ... _the stuff of the world is mind-stuff ... The mind-stuff of the world is something more general than our individual conscious minds._"

James Hopwood Jeans (*1877-1946*), astronomer, was the first to propose that matter is consistently created throughout the "restless" universe, and that the universe has no beginning and no end. In 1943 he wrote *Physics and Philosophy*, wherein he explores reality from two different perspectives: science and philosophy. Jeans stated, "***The universe begins to look like a great thought rather than a great machine***."

Hannas Alfvén (*1908-1946*) pioneered plasma cosmology. Plasma pervades the universe as a gas of positive ions and free electrons with an approximately equal negative and positive charge. The plasma universe is formed and controlled by electricity and magnetism. A lightning storm is an example of partially ionized plasma, and the interior of the sun is an example of fully ionized plasma. Plasma cosmology asserts that because we now see an evolving, changing universe, the universe has always existed and always evolved, and will exist and evolve for an infinite time to come.

Max Planck (*1858-1947*) originated quantum theory and believed that whatever is happening at the subatomic level is the result of consciousness: "***I regard consciousness as fundamental. I regard matter as a derivative from consciousness.*** *We cannot get behind consciousness. Everything that we talk about, everything that we regard as existing, postulates consciousness.*" Planck also said, "*All matter originates and exists only by virtue of a force which brings the particle of an atom to vibration and holds this most minute solar system of the atom together. We must assume behind this force the existence of a conscious and intelligent mind.* ***This mind is the matrix of all matter.***"

Sri Aurobindo (*1872-1950*) was an Indian philosopher who stated, "*All is consciousness – at various levels of its own manifestation … this universe is a gradation of planes of consciousness.*"

Albert Einstein (*1879-1955*) was a theoretical physicist who also influenced the philosophy of science, claiming that "***all of science isn't anything more than the refinement of everyday thinking***." He believed that we participate in a divine plan: "*Whatever there is of God in the universe, it must work itself out and express itself through us.*"

In a letter to his daughter, Einstein wrote: "*There is an extremely powerful force that, so far, science has not found a formal explanation to. It is a force that includes and governs all others, and is even behind any phenomenon operating in the universe and has not yet been identified by us. This universal force is love ... .*"

Einstein developed his special theory of relativity from a thought experiment in which he imagined himself riding a beam of light through space to see how things looked from that perspective. He believed that "*imagination is more important than knowledge*" and also said of himself, "*I believe in intuition and inspiration – at times I feel I am right without knowing the reason.*"

Einstein developed the mass-energy equivalence formula, $E = mc^2$, the most famous equation in the world. He also developed the photon theory of light, examined the photoelectric effect and the thermal properties of light, made contributions to particle theory, and studied the motions of molecules.

Einstein made contributions to quantum mechanics and also voiced two skeptical observations about how it was being interpreted: (1) "*Quantum mechanics is certainly imposing. But an inner voice tells me that it isn't yet the real thing ... at any rate, I am convinced that God does not throw dice*", and (2) "*Whether you can observe a thing or not depends on the theory you use. It is the theory which decides what can be observed.*"

Wilhelm Reich (*1897-1957*) was a medical doctor and psychoanalyst who coined the term "orgone" for a form of biological energy that can be seen as dancing particles if you stare at a blue sky on a sunny day. Reich believed that orgone essentially created all of nature. Although first observed in the human body, he learned through experimentation that ***orgone exists in free form in the atmosphere – and may be the energy that underlies consciousness, life, and cosmic processes***. Orgone is mass-free, present everywhere, is in constant motion -- may be the medium for electromagnetic and gravitational phenomena – may be the energy from which matter is created – and may be responsible for life. Reich further discovered that orgone can be controlled by orgone accumulators, which devices he built to improve the health of his patients.

Wofgang Pauli (*1890-1958*) was a theoretical physicist and one of the pioneers of quantum mechanics. His insight into the limited role that observation plays in experiments is this: *"There remains still in the new kind of theory an objective reality, inasmuch as these theories deny any possibility for the observer to influence the result of a measurement, once the experimental arrangement is chosen."* Pauli also said, philosophically, *"The layman means, when he says, 'reality' that he is speaking of something self-evidently known; whereas to me it seems **the most important and exceedingly difficult task of our time is to work on the construction of a new reality**."*

Erwin Schroedinger (*1887-1961*) was a physicist who developed a number of fundamental results in quantum theory and wave mechanics. He also authored many works in dielectrics, electrodynamics, general relativity, cosmology, and genetics. Schroedinger paid great attention to philosophy, ethics, and theoretical biology.

About consciousness Schroedinger had this to say: *"We don't belong to this material world that science constructs for us. We are not in it; we are outside.   We are only spectators"* … *"**Consciousness is absolutely fundamental. It cannot be accounted for in terms of anything else.**"* … *"The total number of minds in the universe is one. In fact, consciousness is a singularity phasing within all beings."*

Niels Bohr (*1885-1962*) contributed to the understanding of atomic structure and quantum theory. He was also a philosopher and a promoter of scientific research. Bohr proposed that electrons revolve in stable orbits around the nucleus but can jump from one energy level (or orbital) to another. He also conceived complementarity, that objects could be separately analyzed in terms of contradictory properties, such as behaving like a wave when viewed from one perspective and like a stream of particles when viewed from another. Bohr observed that *what happened to one object could influence the behavior of another instantly, even though no physical force connected them*

Bohr stated, "Isolated material particles are abstractions, their properties being definable and observable only through their interaction with other systems." Both this view and complementarity speak of relativity at the macro level – you can only observe and measure the position of something else in relation to the specific position that you are taking for purposes of that observation.

Bohr said, *"There is no quantum world. There is only an abstract quantum physical description,"* and **"Everything we call real is made of things that cannot be regarded as real."**

Werner Heisenberg (*1901-1976*) was a theoretical physicist and a pioneer of quantum mechanics. He formulated the uncertainty principle whereby the more accurately you measure an object's position, the more uncertain you become about its momentum, and vice versa. This discovery led him to believe that **"atoms or elementary particles themselves are not real; they form a world of potentialities or possibilities rather than one of things and facts**."

Heisenberg also made a number of philosophical observations about quantum reality, such as *"The existing scientific concepts cover only a very limited part of reality, and the other part that isn't understood is infinite"* and *"We have to remember that what we observe isn't nature herself, but nature exposed to our method of questioning.*" Heisenberg also believed that consciousness is something science cannot get behind or go beyond.

Richard Feynman (*1918-1988*) was a theoretical physicist who specialized in the behavior of subatomic particles and believed that "*all mass is interaction*" and "*in physics today we have no knowledge of what energy is.*" He also said, "*The electron is a theory we use; it is so useful in understanding the way nature works that we can almost call it real.*"

Feynman came up with a plausible explanation for the two-slit experiment in which particles take different paths to a destination depending on the structure of the experiment. He demonstrated that a particle going from A to B appears simultaneously to take every possible path in spacetime to get there. It is as if the particle does not have a single history but a number of possible histories.

Feynman made some interesting philosophical observations, such as "*a very great deal more truth can become known than can be proven*" and "**the [quantum] paradox is only a conflict between reality and your feeling of what reality ought to be**".

John Stewart Bell (*1928-1990*) was attracted to philosophy but became discouraged by how much philosophers contradicted each other. Instead, he moved to physics where "*you could reasonably come to conclusions.*" Bell specialized in nuclear physics and quantum field theory, in which he formulated Bell's theorem – which states that no physical theory of local hidden variables can ever reproduce all the predictions of quantum mechanics. The implication is that *in addition to known physical forces, there must also be some unknown nonphysical component at work at the sub-atomic level.*

There is also a Bell's inequality which states that no objects with reality and separability can exist in our actual world. This is another way of saying that everything is connected.

Bell believed that quantum theory isn't the whole story, that quantum mechanics reveals the incompleteness of our worldview – and further that quantum mechanics, by its very nature, is resistant to precise formulation. He felt it was likely that *"the <u>new way of seeing things</u> will involve an imaginative leap that will <u>astonish us</u>."* *jumps the groove*

David Bohm (*1917-1992*) was a theoretical physicist who contributed innovative ideas to quantum theory, neuropsychology, and the philosophy of mind. He postulated that *"thought is distributed and non-localized in the way that quantum entities don't readily fit into our conventional model of space and time."* Bohm's main concern was with understanding the nature of reality in general and of consciousness in particular, as a coherent whole. He believed that *"in some sense man is a microcosm of the universe; therefore, what man is, is a clue to the universe."*

Bohm believed that *"deep down the consciousness of mankind is one"*. About universal connectedness he had this to say: *"... the inseparable quantum interconnectedness of the whole universe is the fundamental reality, and that relatively independent behaving parts are merely particular and contingent forms within this whole."*

Bohm deduced **an unseen "quantum force" that guides particles in a way that makes their actions appear random and elusive of precise measurement**. Seeming randomness appears only because we cannot know the precise initial position and velocity of each particle.

Hugh Everett (*1930-1982*) was the physicist who first proposed the many-worlds interpretation of quantum physics, which he called his "relative state" formulation. **He included consciousness as part of the physical universe described by quantum mechanics**. The essence of Everett's theory is that the world we experience is a minute fraction of all worlds and "we" could be existing in many of them. According to this theory, there could be two of you, one in each of two parallel worlds.

Fred Hoyle (*1915-2001*) was an astronomer who agreed that the universe is steadily expanding but rejected the "big bang" theory that he so named on BBC radio, intending this term to be pejorative. To him, the idea that the universe had a beginning is an *"irrational process, and can't be described in scientific terms."* Instead, Hoyle postulated the "steady state" theory in which new matter is continually created as the universe expands, through something he called the "creation field". He believed that continuous creation is no more inexplicable than the appearance of the entire universe from nothing.

Hoyle also promoted panspermia, the idea that life exists throughout the universe, distributed by meteorites, asteroids, or comets. Earth may have been hit by a steady influx of viruses trapped in debris ejected into space after collisions between planets that harbor life.

Hoyle was an outspoken critic of randomness: *"Once we see, however, the probability of life originating at random is so utterly minuscule as to make it absurd, it becomes sensible to think that* **the favorable properties of physics on which life depends are in every respect deliberate** *.... It is, therefore, almost inevitable that our own measure of intelligence must reflect ... higher intelligences ... such a theory is so obvious that one wonders why it isn't widely accepted as being self-evident. The reasons are psychological rather than scientific."*

Francis Harry Crick (*1916-2004*) was a molecular biologist, biophysicist, neuroscientist, and co-discoverer of the structure of the DNA molecule. In the 1970s, he speculated that the production of living systems from molecules may have been a rare event in the universe, but once it had developed, it could be spread by intelligent life forms using space travel technology, a process called "directed panspermia". Crick's later research centered on theoretical neurobiology and attempts to advance **the scientific study of human consciousness**.

Herman Bondi (*1919-2005*) was a mathematician and cosmologist who, with Fred Hoyle and Thomas Gold, in 1948 developed the "steady state" theory of the universe as an alternative to the "big bang". According to this theory, the universe is continually expanding but matter is constantly created to form new stars and galaxies to maintain a constant average density.

Thomas Gold (*1920-2004*) was an astrophysicist and steady state theorist who believe that the universe has no beginning or end. Gold and Bondi hypothesized the spontaneous creation of matter and speculated that within every block of space about 100 meters on a side, there would come into being about one hydrogen atom per year. These atoms gradually condense by their own gravity into large clouds, then into stars and galaxies.

John Wheeler (*1911-2008*), the theoretical physicist who coined the terms "black hole" and "wormhole", believed that ***every item of the physical world has at bottom an immaterial source and explanation***. He talked about "mass without mass" and declared that information is more fundamental in the universe than energy. In a 1940 telephone conversation with Richard Feynmann, Wheeler stated, *"I know why all electrons have the same charge and the same mass … because they are the same electron!"*

    Wheeler believed that this is a participatory universe in that *"we are participators in bringing into being not only the near and the here but the far and long ago"* – suggesting that we are the minds that help manifest the universe.

Hans Peter Dürr (*1929-2014*) was a physicist who made contributions to nuclear and quantum physics, elementary particles and gravitation, epistemology and philosophy. Dürr concluded that, according to quantum physics, everything is included and incorporated into one indivisible potential reality.

Karl H. Pribram (*1919-2015*) was a neurosurgeon and professor of psychology and psychiatry who did pioneering work on the interrelationships of the limbic system, frontal cortex, parietal and temporal lobes, and motor cortex of the brain. Pribram has stated: *"One can no more hope to find consciousness by digging into the brain than one can find gravity by digging into the earth."*

Stephen Hawking (*1942 - 2018*) was a theoretical physicist who made major contributions to cosmology, quantum mechanics, general relativity, and quantum gravity, especially in the context of black holes. With respect to order in the universe Hawking has stated, **"The whole history of science has been the gradual realization that events don't happen in an arbitrary manner, but that they reflect a certain underlying order**, *which may or may not be divinely inspired."* Like Einstein, Hawking was inspired by intuitive moments: *"I rely on intuition a great deal. I try to guess a result, but then I have to prove it."*

Freeman Dyson (*1923 - *) is a theoretical physicist, mathematician and futurist. He believes in a dual origin concept of life: that life first formed cells, then enzymes, and at a much later date formed genes. Dyson says, *"Life may have succeeded against all odds in molding a universe to its purposes."*

Dyson also speaks of a cosmic metaphysics of the mind in which **the universe shows evidence of the operations of mind on three levels**. First level is the elementary physical process in quantum mechanics whereby matter is constantly making choices between alternate possibilities. Second level of mind is the level of direct human experience. Third level is that the universe itself may include a mental component or mental apparatus.

Henry Pierce Stapp (*1928 - *) is a particle physicist who believes that consciousness is fundamental to the universe. Philosophically this is panpsychism, the view that consciousness is a universal feature of all things and the primordial feature from which all others are derived. Two of Stapp's works (*Mind, Matter and Quantum Mechanics*, and *Mindful Universe: Quantum Mechanics and the Participating Observer*) explain his hypothesis of how mind may interact with matter via quantum processes. Stapp believes that **there are two realities, one physical and the other mental**. The physical reality includes the brain. The mental includes consciousness and intentions.

Roger Penrose (*1931 - *) is a mathematical physicist and philosopher of science who has written books on the connection between fundamental physics and human consciousness. He argues that **the laws of physics are inadequate to explain the phenomenon of consciousness**, and that **consciousness transcends logic**. About quantum theory Penrose says, "*It seems to me that we must make a distinction between what is 'objective' and what is 'measurable' in discussing the question of physical reality, according to quantum mechanics.*" The implication is that what is real isn't always measurable.

Penrose also said, " *... almost all the interpretations of quantum mechanics ... depend to some degree on the presence of consciousness for providing the 'observer' that is required for ... the emergence of a classical-like world.*" One of the mysteries of quantum theory is how much the consciousness of the human observer may or may not be contributing to the outcome of experiments.

Seyyed Hossein Nasr (*1933 - *) is an Islamic philosopher who has stated, "*The nature of reality is one other than consciousness.*"

Amit Goswami (*1936 - *) is a theoretical quantum physicist who had vainly been seeking for a description of consciousness within science and realized, "*Instead, what I and others have to look for is a description of science within consciousness.*" Goswami believes in a new paradigm, that "**consciousness is the ground of all being**." He states that in addition to the causal power from elementary particles upward there is also downward causation by consciousness, which is primary.

Fritjof Capra (*1939 -*) is a physicist and systems theorist who has stated: *"The influence of* **modern physics** *goes beyond technology. It* **extends to the realm of thought** *and culture where it has led to a deep revision in man's conception of the universe and his relation to it."*

Capra believes that quantum theory reveals a basic oneness of the universe; and at the subatomic level, *"matter does not exist with certainty at definite places, but rather, shows 'tendencies' to exist."*

Bruce Lipton (*1944 -*) is a cell biologist best known for promoting the idea that genes and DNA can be manipulated by a person's beliefs. Lipton compares the working of a human cell to a personal computer in which the nucleus is *"simply a memory disk, a hard drive containing the DNA programs that encode the production of proteins."* The DNA-containing part of the nucleus can actually be removed without disrupting the normal operations of the cell.

Menas Kafatos (*1945 -*) is a computational physicist and science writer who believes in **a consciousness based reality**. His publications include *The Nonlocal Universe* and *The Conscious Universe*.

Anton Zeilinger (*1945 -*) is a quantum physicist who specializes in the fundamental aspects and applications of quantum information. He believes that experimental metaphysics may lead to explanations beyond quantum theory. Zeilinger has been quoted as saying: *"**I think there is a need for something completely new. Something that is too different, too unexpected, to be accepted as yet**"* and *"This new theory will be so much stranger ... people attacking quantum mechanics now will long to have it back."*

Alain Aspect (*1947 - *) is a physicist whose experiments showed that *a quantum event at one location can affect an event at another location without any obvious mechanism for communication between the two locations*. The time between detection of one photon and the detection of its twin is less time than it would take for light to get from one apparatus to the other.

Eric Lerner (*1947 - *) is a popular science writer and independent plasma researcher. Plasma cosmology holds that the dynamics of ionized gases and plasma play important or even dominant roles in the physics of the universe. In 1991, Lerner wrote the book, The Big Bang Never Happened. In 2014, Lerner and a team of astrophysicists found no difference in the surface brightness (per unit area) of 1,000 near and far galaxies. This is confirming evidence that galaxies are not retreating from the Milky Way (i.e., that the universe is not expanding).

Andrei Linde (*1948 - *) is a theoretical physicist who believes that the universe and consciousness are interconnected: "*Will it not turn out, with the development of science that the study of the universe and the study of consciousness will be inseparably linked, and that ultimate progress in the one will be impossible without progress in the other ...* **will the next important step be the development of a unified approach to our entire world, including the world of consciousness?**"

Donald D. Hoffman (*1955 - *) is a cognitive scientist who believes that "**consciousness and its contents are all that exist**." Two of his papers are *Consciousness is Fundamental* and *Conscious Realism and the Mind-Body Problem*. Hoffman has also written the book, *Visual Intelligence: How We Create What We See.*"

David Chalmers (*1966 - *) is a philosopher and cognitive scientist who believes that **consciousness is a fundamental principle of existence**. He states that it is impossible to explain consciousness in terms of neural correlates in the brain, and that physical theories don't tell us how consciousness arises. Chalmers suggests that a theory of consciousness should take experience as a primary entity alongside mass and spacetime. Chalmers also had this to say about consciousness: *"There's nothing we know about more directly ... but at the same time it's the most mysterious phenomenon in the universe."*

Scientific Speculation

A theory is only as good as its assumptions.

A theory is a supposition attempting to explain something that is not otherwise understood. All theories are tentative in that while they may explain the evidence to date, there is always the possibility that new evidence may emerge which either modifies or refutes the theory in question.

Theories exist only in our minds, to help us interpret or understand some aspect of reality. Good theories are those which both accurately describe current observations and can also predict future observations.

Self-Evident Truths

"Do I exist?" Well, of course you do. If you did not, you would not be able to pose this question. This is an example of a self-evident truth (equivalent to an axiom in geometry). To presume that you don't exist is to commit the logical fallacy of self-exclusion, meaning that you have exempted yourself from your own reasoning.

"Am I conscious?" Well, of course you are. Consciousness is awareness, and you are certainly aware enough both to understand the issue and to pose a question about it. To presume that you are not conscious is another example of the fallacy of self-exclusion.

"Do I have free will?" Well, of course you do. Another self-evident truth. The question assumes that you are able to make and act on your own choices. There were other questions you could have asked, or you could have chosen to remain silent. Determinism, the doctrine that human action is determined by causes external to the will, thus depends on a false assumption. Heredity, environment, and experience may influence the choices you make; but ultimately you are the one who chooses. How you make decisions, whether with awareness or by default, is also your choice.

Logic, the science of reasoning, is the ultimate standard by which all theories need to be judged. One valid reason why a theory cannot be so overrides all the evidence which suggests that it could be so.

A below is a self-evident truth. **B** is an empirical observation. From these two premises, **C** is the logical conclusion.

A. **There is only one reality.** Anything that is unreal does not exist. Everything that exists is real, even if we cannot perceive it with our senses or with instrumentation.

B. **A physical object cannot create itself.** Every physical object, including matter itself, has been created or caused by something other than itself.

C. **The physical universe has a nonphysical cause.** The total of everything which exists in our physical universe was thus created by something which exists beyond the physical realm.

Did early philosophers go through this formal chain of reasoning to reach conclusion **C**? Perhaps not. It may have been something that to them was self-evident. However conclusion **C** may be reached, the following two corollaries necessarily follow:

D. **Reality consists of both a physical aspect and a nonphysical aspect.**

E. **The nonphysical is the primary aspect which created the physical aspect.** Physical reality is thus a manifestation of nonphysical reality.

Definitions

The more precisely we define our terms, the more clarity we bring to the subject in question. No conclusion can be valid if it contradicts the definition of a concept upon which the argument is based.

If we define "universe" to mean "all existing things", then we must conclude that the universe has no shape, no boundaries, and no edges – because to have a boundary implies that something else exists on the other side of that supposed dividing line. If something were to exist on the other side of the universe's (imagined) edge, then by definition, that other something is also part of the universe. Thus, to conclude that the universe has a shape is to commit the fallacy of self-exclusion – i.e., the universe (everything which exists) can have a shape only if something other than the universe also exists.

Similarly, the concept of size does not apply to the universe – because to have a size also means having a shape. If it is impossible to ascribe a shape or size to the total of everything which exists, then we must conclude that the universe is limitless, boundless, infinite. This is mind boggling, to say the least. Nevertheless, it must be so.

Dichotomies

It is common practice to divide competing ideas into two sharply defined categories, then argue for one against the other in an adversarial way. The logic is thus: If theory A is right, then theory B must be wrong. This methodology works only when A and B are the only two mutually exclusive possibilities – like saying that a color is either black or not black – or deciding in court that one is either guilty or not guilty. This is sound reasoning only if B is defined as including every possibility that A does not – and vice versa.

There are two competing theories to explain the creation of the universe. This is a genuine dichotomy. If A is true, then B must be false – and vice versa.
   A. Singularity theory: the universe spontaneously created itself by means of a "big bang" or some other one-time event.
   B. Infinite universe theory: the universe has always existed and will always continue to exist.

Below is an example of a false dichotomy. It is a mistake to assume that if **A** is false, then **B** must be true. Both **A** and **B** can be false, because there is a third possibility – namely that matter in the infinite universe could be continuously created in ways that change the density of matter through time.
   A. Steady state theory: matter in the infinite, expanding universe is continuously being created at a rate that maintains all matter in the universe at a constant density.
   B. Big bang theory: the universe was created at a finite point in time.

Another false dichotomy concerns the origins of species that has for centuries fueled the debate that one either believes in creation or evolution. Many believe in both.

A. Creationism: plants, animals, and humans were created exactly as they are.
B. Evolution: plants, animals, and humans evolved on their own

Creationism versus evolution does not include all possibilities. Evolution itself could have been created.

One can deny that (an assumed) God had a direct hand in creating new species and also believe that God created evolution. Creationists believe in a literal interpretation of the Bible, whereas there are many proponents of evolution who interpret the Book of Genesis as a metaphor rather than fact.

Perhaps creation and evolution are intertwined. If evolution began from a single species, that species somehow must have been created – because no prior species existed from which it could have evolved. If one original species was created, then why not more than one? Is it possible that entirely new ancestral lines may have been created from time to time?

Probability

Nothing ever happens by chance. Nature isn't random. It only appears so when we don't understand what caused the effects we are observing. We use probability to predict events or outcomes when we don't know (or cannot measure) what caused them.

The outcome of a coin toss is not random. There are precise forces which govern exactly how many revolutions a coin will make over a given trajectory. If someone were willing to make the necessary calculations, it would be possible to construct a machine with settings to throw either a head or a tail at will, every time. Without such a machine and knowing that there are only two possible outcomes, we guesstimate a probability factor of 50:50 to predict how many heads (or tails) we are likely to get over many tosses. However, this probability factor tells us absolutely nothing about the next toss, the outcome of which is completely independent of all previous tosses.

At the quantum level, there is seeming randomness only because we don't know the precise initial position and velocity of each particle or wave. Just as in a coin toss, Stephen Hawking tells us, "*Quantum theory does not predict a single definite result for an observation. Instead, it predicts a number of different possible outcomes and tells us how likely each of these is.*" Likelihood, however, tells us nothing about cause and effect.

Hawking and others also state (with compelling argument) that in quantum mechanics, "*An object has not just a single history but all possible histories*". That much is clear. What is unclear is the supposed explanation that the probabilities for various possible histories supposedly cancel each other out to reveal the one actual history for the object in question. Probabilities are guesses, not causes. The probability of throwing a "12" with a pair of six-sided dice is 1 in 20. Whenever a "12" is thrown, does that mean that 19 other probabilities cancelled themselves out? No. There were precise forces that determined the actual paths that the two dice took instead of perhaps hundreds of other paths that would have yielded 10 different possible outcomes.

About 66 million years ago, an asteroid crashed into Earth, the repercussions from which killed off the dinosaurs. Land dwelling creatures which survived the initial impact starved to death because clouds of debris blocked the sun worldwide, thus halting photosynthesis. Do asteroids crash into planets at random? No. There are precise forces that kick asteroids out of orbit and guide the trajectories that these rocky bodies take. Cataclysmic events such as these only seem random to us because we imagine many other possible outcomes had the physical forces been slightly different.

For sake of argument, let us suppose that the asteroid which killed the dinosaurs could have taken, say, 100 possible paths. It is tempting, then, to speculate that there was a probability of only one percent that the dinosaurs would be wiped out. Probability, however, is a speculative mind game that has nothing to do with causality. Whether there are 99 theoretically possible other outcomes (as in the asteroid conjecture) or only one possible other outcome (as in a coin toss), in both cases the principles of physics ultimately determine what happens.

Fred Hoyle used this metaphor to make his position clear about randomness: *"A junkyard contains all of the bits and pieces of a Boeing 747, dismembered and in disarray. A whirlwind happens to blow through the yard. What is the chance after its passing that a fully assembled 747, ready to fly, will be found standing there? So small as to be negligible, even if a tornado were to blow through enough junkyards to fill the whole universe."*

Another metaphor used by mathematical physicist Fred Hoyle is that of a blind man unscrambling the colored faces of a Rubik's cube. Because this man cannot see whether any twist brings him closer or farther from his goal of ordering the sections of the cube, he has to work by random trial and error – which moves, at the rate of one per second, would take 126 billion years to complete the task. It is, therefore, highly unlikely that living species could have come about by random mutation and natural selection.

"God does not play dice with the universe" is how Einstein dismissed randomness. By this remark he was not implying the existence of a personalized deity, but rather *"behind all the discernible laws and connections, there remains something subtle, intangible, and inexplicable."*

Later in life Einstein remarked, *"The idea of a personal God is alien to me and seems even naive."*

Assumptions

Assumptions limit our ability to experience. Once we believe a certain theory to be true, we tend to overlook or downplay evidence that conflicts with it. It is easier to explain away anomalies in terms of what we know, rather than to use them as opportunities to discover what we do not know.

If we assume that the entire universe began from a big bang, then it logically also follows that (a) this was the only moment of creation, (b) the universe was at that time imbued with a total fixed amount of energy, and (c) the universe is a closed system into which no more energy is ever introduced. Many scientists assume that the energy of the universe must be a fixed constant in order to provide stability, to make it so that things don't just appear from nothing. This belief closes the door to the possibility that new things could appear, not from nothing, but from a nonphysical source. Ironically, it may be that a continuous input of nonphysical energy is required for the very purpose of providing stability for the cosmos, for the atom, and for life.

From the assumption (c) that the universe is a closed system, it logically follows (d) that all observed energetic phenomena must have physical causes that have been present since the supposed big bang. This leads cosmologists to speculate that there must be some hidden dark matter causing otherwise unexplained gravitational effects on galaxies and stars – and quantum theorists to speculate that there must be some hidden particles causing otherwise unexplained electromagnetic effects on other subatomic particles. The possibility that additional energy could be introduced into outer or inner space from nonphysical sources has been ruled out presumptively.

The above limiting assumptions are all that stand in the way of developing a viable theory of the origin of everything. If we instead assume that (1) the physical universe has a nonphysical cause, and (2) the nonphysical continuously interfaces with the physical, then an entirely different and consistent picture unfolds across all fields of scientific endeavor. All the pieces of the puzzle can fit, so to speak – without the need for hypothetical (assumed) virtual particles, antiparticles, antimatter, antigravity, dark matter, or dark energy.

Bang Goes the Theory

The universe behaves more like a mind than a machine.

There is a prevailing notion that the universe was created by a "big bang" explosion that happened some 13.8 billion years ago. This date was arrived at by working backwards in time from equations that supposedly measure the universe's rate of expansion.

According to this theory, the entire universe began from a single point (or micro-dot) violently exploding out pure energy which almost instantly became particles and then atoms, which combined to form elements, molecules, gases, stars, and galaxies. In other words, the universe spontaneously created itself. This is mythology, not science.

The big bang theory is the prevailing and firmly entrenched cosmological model for the universe – and it is entirely beyond the realm of possibility. The immortal words of Will Rogers come to mind: *"It isn't what we don't know that gives us trouble. It is what we know that ain't so."*

If the entire universe began from a single point, one is tempted to ask how this point was created. To seek a cause for a point, however, is a meaningless pursuit because points are a convenient fiction. They do not exist.

<u>First question</u>: how can a point explode? Points are mathematical abstractions used to specify locations on a graph. A point, by definition, has only position but zero size and zero mass. Points, as such, do not exist; they are nonphysical. To claim that the universe began from a single point or dot is to infer (inadvertently, but with impeccable logic) that everything which is

physical ultimately has a nonphysical origin. Some theorists quibble that it was neither a point nor a dot that exploded but something of zero dimensions into which had been previously compressed all the matter of the universe. Same difference. To have zero dimensions is to have zero existence.

Second question: where was this point allegedly located? Location refers to the relative positions of objects with respect to each other. You can only locate something by measuring how far it is from something else – but if nothing at all exists, no such measurement is possible. In the total absence of any physical objects, the concept of position/location has no meaning. Locations are within the universe. The universe itself has no location. Space is in the universe; the universe is not in space.

Third question: how is it possible to affix a point in time at which the universe began? Time measures the rate of change of the relative positions of physical objects (i.e., within the universe). Without material objects moving about, time has no meaning. There is no form of existence which predates existence. Time is in the universe; the universe is not in time.

The assumption that the universe had a beginning (i.e., in time) defies logic. "Beginning of time" is a contradiction in terms. There can be no point in time at which time began. We cannot even say that the universe has always existed, because "always" implies an external time factor. All we can logically say is that the universe is.

The universe has no beginning in time. By the same reasoning, it cannot have an ending in time – because that would be to suggest that there could be a point in time at which time ceased to exist. The universe has no beginning and no end. The universe is infinite.

Flawed Science

1908: Henrietta Swan Leavitt discovered that very bright variable stars (Cepheid stars) pulsate in predictable ways. They obey a period-luminosity relationship: the longer the period of a Cepheid's fluctuation, the more intrinsically luminous it is. Leavitt erred in assuming that this luminosity provides a predictable way to measure distance to these stars. This period-luminosity relationship, however, is a function of the life cycle of physical changes within each individual star and has nothing to do with its distance from Earth. An older Cepheid star with slower pulsations in a nearby galaxy would thus appear to be further away than a newer Cepheid star in a distant galaxy.

1925: Vesto Slipher observed that light from spiral nebulae is "redshifted", meaning that its frequency drops towards the red end of the visible spectrum and its wavelength simultaneously increases by a corresponding amount. Slipher speculated that this is consistent with a light source moving away from the observer and somehow "stretching" the wavelength of the light it emits – and jumped to the erroneous conclusion that most clusters of stars are moving away from us. The misconception that an increase in wavelength can happen only if the light source is moving away comes from confusion with the Doppler effect. Slipher's error of mistaking redshift for Doppler is science's biggest blunder of all time – upon which error are based the "big bang", dark matter, and dark energy myths.

In a Doppler effect, the frequency of **longitudinal sound waves** travelling through an elastic medium (air) remains constant; but if the source is in motion towards the observer, these uniform sound waves bunch together, creating **the illusion of an increase in pitch/frequency** to the observer's ear. (Conversely, if the source of the sound is in motion away from the observer, the illusion of a lowering in pitch/frequency is created.) In a Doppler effect, both the frequency and wavelength of the emitted sound remain constant. However, if the source of the sound is in motion, a distortion in frequency occurs at the point of observation.

The classic example of the Doppler effect is the approach and recession of an emergency vehicle. As an ambulance approaches, the pitch of its siren appears to become higher – because sound waves of the same exact length hit your ear more frequently (i.e., bunch together), creating a distorted increase in pitch/frequency. As the ambulance moves away, sound waves of the same exact length hit your ear less frequently, creating a distorted lowering in pitch.

Doppler radar bounces a microwave signal off a moving target and analyzes how this object's motion distorts the frequency of the returned signal. Velocity of the object is inferred as a function of the difference between the distorted frequency and the actual frequency.

In a redshift, **transverse light waves** travelling through space from a stationary source simultaneously decrease in frequency and increase in wavelength over extreme distances. In a redshift, there is **an actual increase in wavelength**. In Doppler, **wavelength is constant** and there is a distortion in the perception of the frequency (which is also constant).

For over 90 years, theoretical physicists have been falsely assuming (a) that galaxies are in motion away from Earth, and falsely presuming (b) that the alleged velocities of said galaxies is proportional to the degree by which light emitted from them is shifted towards the red end of the spectrum. **Redshift is a reliable measure only of distance and nothing else.** The farther away a galaxy is, the more its light shifts to the red end of the spectrum. That is all a redshift can tell us. Nothing more.

1925: Alexander Friedmann, a mathematician, proposed that the universe could be either expanding, contracting or remaining static. Friedmann developed equations to predict either the rate of expansion or rate of contraction, once it was known which was the case. Subsequent theorists assumed that the universe was expanding and used the Friedmann equations to prove their point.

1927: George LeMaitre, an astronomer, on his own developed the same equations as Friedmann. LeMaitre, however, was convinced that the universe was expanding and so he proposed his "theory of the primeval atom" to explain why he thought it was expanding.

1927: Edwin Hubble theorized that galaxies are receding from the Milky Way and the farther they are away, the faster they are receding. Hubble's conclusions, however, were based on (a) the Leavitt error of assuming the brightness of a Cepheid star is a function of its distance, (b) the Slipher error of assuming the redshift of a star cluster is a function of its velocity away from the observer, and (c) some mathematical errors of his own that may have been contrived to justify his foregone conclusion.

1929: Edwin Hubble published *A Relation Between Distance and Radial Velocity among Extra-Galactic Nebulae*, a paper in which he estimated the "radial" velocities of 46 star clusters, on the assumption that these nebulae were travelling on straight line paths diverging from the center of an explosion. Then, from their radial velocities, he estimated the distances of only 24 of these nebulae. Hubble (a) assumed that redshift is a measure of velocity away from the observer, (b) used trigonometry to estimate presumed velocities away from a central explosion, (c) from these radial velocities inferred the distances of a select 24 of these nebulae, and (d) plotted

22 of these results on a graph relating velocity to distance, five of which points coincided with a straight-line relationship between these two variables.

1929: From the above paper, Edwin Hubble and Milton Humason formulated Hubble's law, which states that objects in deep space have a presumed relative velocity away from Earth, and their velocity of recession is proportional to their distance from Earth.

1930: Richard Tolman devised a surface brightness test to determine whether the universe is static or expanding. The Tolman test compares the surface brightness of galaxies to their degree of redshift. In a static universe, the surface brightness (light received per unit area) of a galaxy is constant regardless of its distance (and regardless of its redshift). In an expanding universe, surface brightness diminishes with distance. For 88 years, mainstream astrophysicists have never checked the validity of their assumptions by means of the Tolman test. They all accept on blind faith Slipher's error of mistaking redshift for a Doppler effect.

1931: George LeMaitre published the English version of his earlier paper now entitled, *"A homogenous Universe of constant mass and growing radius accounting for the radial velocity of extragalactic nebulae."* LeMaitre initially called his theory the "hypothesis of the primeval atom" and described it as the "cosmic Egg exploding at the moment of creation." In addition to being an astronomer, LeMaitre was also a Catholic priest who felt comfortable with the notion that God had created the atom/egg that subsequently blew up to create the universe.

1949: Fred Hoyle, an opponent of the primeval atom theory, sarcastically called it the "big bang" on a BBC radio broadcast. The name stuck.

1964: Cosmic microwave background (CMB) radiation was discovered. Big bang proponents had been searching for confirming evidence for their singularity theory, and this appeared to be it. CMB radiation is mistakenly believed to be thermal radiation left over from "recombination", the epoch during which charged electrons and protons supposedly first became bound to form electrically neutral hydrogen atoms. The assumption is that hydrogen, the lightest and simplest element, was made exclusively during the big bang. Hydrogen, however, is the most abundant element permeating the entire universe, and it may even be continually created.

CMB radiation looks structurally the same in every direction. This is consistent with the logic that tells us that the universe did not begin anywhere and isn't located anywhere. Space is in the universe. The universe is not in space.

NASA confirms that the CMB follows the precise curve for blackbody radiation. A blackbody is an opaque object in space that absorbs radiation of all wavelengths that falls on it. Then, when the blackbody is at a very hot and uniform temperature, it emits radiation that is outside the visible spectrum of light. NASA's measurements show that this blackbody curve peaks at 0.3 cm. wavelength and 100 GHz frequency, which is at the high end of the microwave spectrum. The blackbodies in question could simply be interstellar dust.

The cosmic microwave background is smooth and looks the same in all directions for the same reason that a fog looks smooth and uniform in all directions. The CMB is, in a sense, an electromagnetic fog.

2014: Eric Lerner and a team of astrophysicists applied the Tolman test by measuring the surface brightness (per unit area) of over 1,000 near and far galaxies. They found that in every case, surface brightness remains constant regardless of distance. If any galaxy had been in motion away from us, its surface brightness would have been much less than that of the other galaxies measured. Conclusion: there is zero tangible evidence that galaxies are moving apart and overwhelming evidence that they are not. [*International Journal of Modern Physics D vol. 23, No. 6 (2014) 1450058*]

From LeMaitre's flawed theory and Hubble's flawed measurements, theorists jumped to the conclusion that galaxies are being propelled apart by a single explosion that spontaneously increased the matter in the universe from zero to whatever it is now. Presupposing this foregone conclusion to be true, scientists believed that by tracing back in time they would see a universe that was progressively diminishing in size, until eventually they could pinpoint a "big bang" explosion that created the universe.

What the Hubble space telescope views is a spherical horizon, some 13.4 billion light-years away. What lies beyond that horizon, no one knows. This is analogous to someone on a ship at sea, with no land in sight, believing that the ocean is circular and extends only as far as he can see through binoculars.

If our spherical horizon has a radius of 13.4 billion light-years, then its diameter is 26.8 billion light-years. In other words, light emitted by galaxies on one edge of our horizon travels for 26.8 billion years to reach galaxies at the extreme opposite edge of our horizon. This is 13 billion years longer than big bang theory tells us the universe has existed. We also have no way of knowing how far the universe extends beyond our horizon.

A Child's Perspective

Parent: "***Once upon a time, a teeny weeny dot exploded, creating everything that exists.***"

Child: "***Who made this dot?***"

P: "***Nobody. It was just there.***"

C: "***Where? If nothing existed, there was no place to put a dot.***"

P: *"Stop interrupting. I am trying to tell a story."*

C: *"And how could a dot exist before there was such a thing as existence?"*

P: *"Never mind. It just did."*

C: *"When did this happen?"*

P: *"Almost 14 billion years ago."*

C: *"A year is the time it takes for the Earth to circle around the sun, isn't it?"*

P: *"Yes."*

C: *"Before there were planets or suns, there was no such thing as years. Correct?*

P: *"Yes."*

C: *"So how can you say this story began once upon a time? If there weren't any years, there wasn't any time."*

P: *"Stop trying to be so logical. Not everything is logical."*

C: *"Apparently not. So why should I believe this story?"*

P: *"Because I said so."*

An Open System

The false assumption that the universe was created by a big bang or other singularity leads to a false conclusion: that all the energy and matter in the universe was ultimately injected into it on day one – meaning that the universe is a closed system into which no more energy can ever be admitted.

The logical errors subsumed in the big bang theory are the result of attempting to find an external cause for a system within that system. If the universe has an external cause, this cannot be understood by a reverse chronological examination of its internal effects. True understanding can only come from starting with a known (or plausible) nonphysical cause, then moving forward to explain its physical effects.

When we sweep away the fallacies, a very different picture of the universe emerges. To be without beginning or end is to be infinite. To be without limits in space is to be infinite. As mind-boggling this concept may be, the universe must in fact be infinite.

The universe has no center because it has no shape. To have a shape would mean that the universe would have to have a boundary beyond which something else must exit – but if anything else also existed, it would by definition be part of the universe (of all existing things).

The only way in which a big bang could make any sense is as a metaphor for a singular "explosion" of nonphysical energy into physical reality, an event that could only have been orchestrated by nonphysical reality. If nonphysical energy can create a singular creative event, then it can just as easily create a continuing series of such events.

The above logic leads us to the inescapable conclusions that the infinite universe has no beginning, no location, no boundaries, no shape, and no limits. The only limits are in our understanding. The universe itself is limitless – without location, without beginning, and without end.

When galaxies were supposedly observed to be moving away from each other, the following assumptions were made.
1. the visible/known universe is all there is [*false*],
2. the universe is expanding [*unsubstantiated*],
3. this expansion was caused by a super colossal explosion [*false*],

4. if galaxies are moving away from each other this also means they are moving away from the site of said explosion [*false*],
5. the rate of expansion is mathematically related to the time since the hypothetical causative explosion [*false*], and
6. the universe approximates a sphere in shape [*false*].

According to these assumptions, (a) the universe is 13.8 billion years old, and (b) there was a moment of extremely rapid inflation that defied the laws of physics. Supposedly, all the matter that eventually became hundreds of billions of galaxies exploded into space with an initial velocity many times faster than the speed of light.

If you start with the observed steady rate of expansion and work back in time, there comes a point where the mathematics falls apart. At 13.8 billion years ago, the evidence indicates that there was much more matter/energy in the universe than the math predicted. Rather than concede that said matter must have predated the big bang, it was explained as evidence that the big bang had been more powerful than anticipated, hurling projectiles faster than light speed - a physical impossibility. A mathematical anomaly which disproves the theory was incorporated into the theory. Interesting.

If nonphysical energy is powerful enough to create the universe with a big bang, then it is also powerful enough to create the universe by other means. Logic tell us that a big bang could never have happened. Therefore, the universe must have been created from the nonphysical realm in some other way – or in multiple ways. Perhaps Fred Hoyle was right in as much as the universe is being continually created.

The physical world in which we are immersed is the only frame of reference we have from which to try to understand everything else in the universe. Our senses tell us that everything around us is physical, including both solid objects and invisible radiation that we can detect with instrumentation. It is difficult to grasp the idea that everything in our physical world could have a nonphysical origin. We have no idea what a nonphysical reality could be like, nor do we have the language to describe it. Thus, our first response to any uncertain phenomenon is to look for something physical as its cause.

Logic tells us that something cannot be created from nothing, that everything which exists must have been caused by something else which also exists. This is a self-evident truth. However, our physically oriented minds overlook the possibility that the "something else" in question could be nonphysical. Our penchant for seeking physical causes for everything has led to hypothesizing some questionable theories – including those of dark matter, dark energy, Higgs theory, and antimatter – all of which depend on two dubious assumptions: (1) that the universe was created by a single explosion, and (2) that all of the energy and matter in the universe was imparted to it by that single moment of creation. In other words, the universe is supposedly a closed system containing a fixed amount of energy.

Dark Matter Myth

Dark matter is hypothetical matter that supposedly fills the dark spaces between stars and galaxies and is inferred to exist because of the presumed gravitational pull it appears to have on visible matter. Measurements of its gravitational effects on galaxies suggest that dark matter accounts for some 85 percent of the matter of the universe and as such is believed to be the force that holds the universe together.

Dark matter cannot be seen by telescopes, nor detected by any other means. Light passes right through dark matter, which neither emits nor absorbs light nor any other electromagnetic energy. Dark matter does not interact with normal matter and does not participate in nuclear fusion. Dark matter has none of the properties of matter. Dark matter does not have any properties at all – because dark matter does not exist.

Dark matter was hypothesized (imagined) to explain the gravitational pull on galaxies that is supposedly keeping the universe from expanding too quickly. Although galaxies are increasing in size through time, they are not moving apart from each other. The universe is not expanding. Therefore, there is no gravitational force controlling this non-expansion – and no reason to postulate (assume) dark matter.

Dark Energy Myth

Dark energy is a hypothetical (presumed) energy source that can neither be detected nor measured. Its existence is inferred by observable effects on the movement of galaxies.

The gravitational effect of dark matter was originally believed to be slowing down the expansion of the universe. When calculations suggested that the universe may be expanding at an accelerating rate, then dark energy was postulated (assumed) to be an unseen force opposing dark matter, thereby reducing its effects.

The universe is not expanding. There are no mysterious forces playing tug-of-war with its rate of non-expansion – no dark matter, and no dark energy opposing dark matter.

Higgs Theory

There is an elaborate theory that attempts to explain how pure energy released from the big bang first became a singular kind of fundamental particle, which then immediately deteriorated into elementary particles. Higgs theory postulates that there is a field or grid of energy spread throughout the universe, and "massless" particles receive mass by passing through it. This theory is predicated on the assumption that the universe was created by a big bang.

"Particle" literally means "a tiny bit of matter." Thus, a massless particle is a contradiction in terms. In the context of this theory, it would be more accurate to use the term "pre-physical particle", suggesting a tiny quantum of energy with the potential to be converted into matter.

Somehow the energy within the hypothetical universal grid interacts with the energy of the pre-physical particle to produce a temporary mass-bearing particle called the "Higgs-boson", whose existence is only inferred and said to last for a billionth of a second before it deteriorates into various elementary particles that form the building blocks of our physical reality. The Higgs-boson is thus attributed by some as being either (a) the "god particle" that created everything else, or (b) the intermediate step through which all of creation happened.

This theory is a convoluted attempt to trace everything in the universe back to a single causal moment – and it is inconsistent. It falsely assumes:
    (a) the big bang first created only Higgs-bosons, and
    (b) a universal grid must have pre-existed the big bang.

In 2012, after hundreds of millions of failed collision experiments, on such collision apparently produced elementary particles believed to be the result of the instantaneous decay of one Higgs-boson. A success rate of less than

0.000001 percent suggests that this result was most probably due to experimental error. Since this 2012 experiment, Higgs theory has drawn a blank.

For this theory to be viable there must have been three separate acts of creation: (1) a big bang, (2) the Higgs field, and (3) the programming of the Higgs-boson to become the ancestor of all other particles – all three of which acts require a nonphysical explanation.

Antimatter

Antimatter supposedly consists of antiparticles which have the same mass as particles of ordinary matter but opposite electromagnetic properties. If particles meet antiparticles, they annihilate each other, leaving pure energy. According to this theory, most of the matter which has ever existed has been consumed by antimatter and what has survived is only an infinitesimally tiny fraction of that which was originally created.

Question: do all particles have electromagnetic charges? If not, then neutral particles must be exempt from antimatter annihilation theory.

The antimatter hypothesis rests on the questionable assumption that causal events are random, that incredibly huge amounts of matter go flying around in space until it collides with similarly huge amounts of its nemesis antimatter, drawn together by electromagnetic attraction. Somehow, the humungous total matter we know as the universe miraculously survived these random and fateful collisions.

For every particle of matter created there is supposedly a corresponding particle of antimatter also created. If this were true, then there could be no survivors of this cosmic battle. Every particle of matter would ultimately be eliminated by its evil twin, so to speak. The only way any actual particles could survive is if their corresponding antiparticles did not exist. This is the logical fallacy of self-exclusion. The conclusion can be true only if the premise is false.

The antimatter hypothesis goes against the principle of conservation of energy, which states that energy can be neither created nor destroyed, it can only be transformed. This is another way of saying that energy is never wasted. The nonphysical source of matter would not have indulged in such an incredible waste of energy as to create unbelievably huge quantities of matter/antimatter, the sole purpose of which would be to eliminate itself.

Deliberate Intention

A big bang is a primal event that can only be explained by nonphysical energy transforming itself into matter. Whether this process happens once or continuously – whether it happens directly or through an intermediate stage – whether it happens instantly or after a lapse of a billionth of second – the result is the same. The nonphysical becomes physical.

If a Higgs field exists, it would have had to have been created deliberately; it could not have been a random event – and it would have had to have been created prior to the alleged big bang.

A big bang could not have happened at random. If it happened at all, it would have taken intricate planning and orchestration to produce a universe – the same intricate planning and orchestration required for continuous creation.

Gravity, electromagnetism and nuclear attraction must have been precisely calculated and coordinated to enable the universe to develop and expand in the ways that it does, and especially to allow for the evolution of life forms, including ourselves. If even one of these forces had been miscalculated by a tiny fraction, neither we nor the universe could exist. It appears that the nonphysical reality which set the universe into motion also created the laws of physics that we use to try to understand the universe.

Both a big bang and continuous creation require both intention and planning. Therefore, a thought process must be responsible. Perhaps nonphysical reality is a realm of pure thought. However it happens, the inescapable conclusion we are left with is that thoughts create reality.

A Tale of Two Theories

Sometimes opposing theories are both wrong.

There are only two ways the universe could have come into existence – either (a) by a singular event, or (b) continuously. There is no third possibility.

If the universe was created by a singularity, then it has a finite existence. If it was not created by a singularity, then the universe is infinite. Again, no third possibility.

In 1927, George LeMaitre developed the singularity theory. He claimed to have discovered that the universe is expanding – and assumed the universe to be of constant mass but with a growing radius. To account for the supposed radial velocity of extragalactic nebulae, LeMaitre postulated the "primeval atom" (or "cosmic egg") theory whereby the universe began with a single quantum that exploded.

In 1948, Fred Hoyle postulated the "infinite universe" theory in which new matter is continually created as the universe expands – with new galaxies developing to fill in the spaces left by the galaxies that are moving further apart. Hoyle believed that continuous creation is no more inexplicable than the appearance of the entire universe from nothing, although it had to be done on a regular basis.

The stage was set for a winner-take-all debate. If LeMaitre was right, then Hoyle had to be wrong – and vice versa – or so it would seem. Both competing theories were flawed, however. It was a false dichotomy. There was a third possibility that had been overlooked.

Ironically, it was Fred Hoyle who coined the term, "big bang", on a 1949 BBC radio show, intending it to be insulting to primeval atom theorists. To Hoyle, the idea that the universe had a beginning is an *"irrational process, and can't be described in scientific terms."*

Hoyle's logic was flawless, in two ways: (1) The universe cannot have had a beginning in time -- because time is in the universe, the universe is not in time. (2) If the universe has no beginning, then it must be infinite.

Hoyle's version of infinite universe theory, however, included a "steady state" caveat in which it was believed that the density of matter in the expanding universe remains unchanged due to a continuous creation of new matter – that the universe is always expanding but maintains a constant average density – i.e., new stars and galaxies are being created at the same rate that old ones become unobservable. In other words, Hoyle lumped together two different concepts into the same theory: (1) the universe is continually being created, and (2) the universe is in a steady state of uniform density.

According to steady state theory, the distribution of matter was expected to be homogeneous and have the same physical properties when viewed on a large enough scale. This is not what telescopes reveal, however. As they pick up images that have been sent from the distant past, galaxies are much smaller in size – and no new galaxies have been seen to fill the spaces between galaxies that appear to be moving apart. The density of the universe has apparently been increasing through time. This evidence challenges steady state theory, but not continuous creation theory. To reject both parts of the infinite universe theory because one is flawed is like throwing the baby out with the bath water, metaphorically speaking.

Big bang proponents have taken failure of steady state theory to be proof of their singularity theory. Steady state and big bang are not the only options, however. It is possible that the infinite universe could be continually created in a way that has nothing to do with the steady state assumption.

Both theories agree that the universe is expanding. They differ in (1) what is causing that expansion, and (2) the pattern this expansion is following. Big bang theory assumes the universe is a finite and expanding sphere, whose radius from the explosion is continually increasing. Steady state theory assumes that the limitless universe is expanding without any geometrical restrictions.

Both theories are flawed. The big bang theory assumes (illogically) that there was a moment in time at which time began. The steady state theory assumes (without evidence) that new matter is continually created at a rate that enables the universe to maintain a constant average density through time.

Modern space telescopes pick up images of how far distant galaxies looked billions of years ago. They were much smaller in size than nearby galaxies appear now. The universe has apparently increased in density through time; therefore, the steady state theory is wrong. That does not make the big bang theory right, however.

Big bang (A) versus steady state (B) is a false dichotomy – because A and B are not mutually exclusive. There is a third possibility: continuous creation (C) at a rate that enables the universe to increase in density.

Expanding or Infinite

If the universe is infinite, then expansion of its supposed boundaries is a contradiction in terms.

Both big bang theory (A) and steady state theory (B) depend on the assumption that the universe is expanding. Infinite universe theory (C), however, does not. If the universe is infinite, then expansion is a non-issue. Into what space could something without boundaries be expanding? How can something of limitless volume be increasing in size? How could something of limitless mass be getting bigger? Infinity plus 100 billion galaxies is still infinity.

A 2014 study by Eric Lerner and a team of astrophysicists suggests that the universe is not expanding. The big bang theory suggests that in an expanding universe, objects should appear fainter but bigger as they move away. Their surface brightness (brightness per unit area) should decrease with distance. Thus, in an expanding universe, the most distant galaxies should have hundreds of times dimmer surface brightness than similar nearby galaxies, making them undetectable with present-day telescopes.

This is not what observations reveal, however. As similar objects get farther away (or seem to), they appear fainter and smaller – but their surface brightness (per unit area) remains constant.

Lerner's team carefully compared the size and brightness of about 1,000 nearby and extremely distant spiral galaxies, matching the average luminosity of the near and far samples. They found the surface brightness of near and far galaxies to be identical. These results are consistent with ordinary geometry if the universe is not expanding – and contradict the diminishing of brightness that would be consistent with an expanding universe.

If Lerner's results are accurate, then something is seriously wrong with the measurements which suggested to LeMaitre, and Hubble that the universe is expanding. Hubble's Law states that galaxies are moving away from the Milky Way at a velocity proportional to their distance. Hubble's Constant is the unit of measurement used to describe the rate of expansion of the universe.

A "constant" in mathematics is a component of a relationship between variables that does not change its value. A constant has nothing to do with causality, however. It is whatever contrived number is required to make the left side of the equation consistently match the right side. In other words, it is a value "plugged" into an equation.

It turns out, however, that the Hubble Constant is really a variable. In 1929, Edwin Hubble estimated its value to be 160 km/sec/million light-years. In 1956, its value had dropped to 58 km/sec/Mly – in 1958 to 24 km/sec/Mly – in the early 1970s to 18 km/sec/Mly – and currently it is back up to 23 km/sec/Mly.

Expansion theory suggests that galaxies become bigger as they move further away, on the assumption that the size of galaxy is a function of its age. But if galaxies are not in fact moving, then their age and size are unrelated to their distance - and the only variable is the farther away they are, the smaller they appear to be. An observer expecting to see evidence of expansion based on the relative size of a galaxy would thus tend to overstate its distance – a galaxy of given size would appear to be much further away than it really is. In other words, expectation tends to exaggerate distance calculations.

If galaxies are not moving apart, then the universe is not expanding; and there is no way to trace back in time to the "big bang" that never happened. The universe is infinite.

Continuous Creation

The universe is constantly evolving.

In ancient Egyptian and Mayan cosmology, the center of the Milky Way galaxy is the place where stars are born. We now know that at the heart of every spiral galaxy there is a super massive black hole whose gravitational pull imparts a spin to that galaxy, giving it a disk like shape. These black holes also spew out pure energy that becomes particles, atoms, molecules, gases, and eventually stars.

When a star dies, it creates a black hole that gives off an explosive interstellar jet-stream of dust, hydrogen, helium, and other ionized gases – the materials from which new stars are created. We are witnessing a stellar cycle of life: the death of stars becomes the birth of new stars – a kind of cosmic recycling.

> **Thought experiment**: Imagine that you could view the cosmos from the perspective of nonphysical reality, where time does not exist. Whatever takes eons to happen in physical reality appears to be happening instantly – the ultimate in fast forwarding. You would see stars exploding, sending off multiple stars that explode into multiple stars, that explode into multiple stars, and so on -- the most super colossal fireworks display imaginable.

Evolution of Galaxies

Galaxies come in three basic shapes: irregular, elliptical, and spiral:

- <u>Irregular shaped galaxies</u> are often chaotic in appearance, have no discernable structure, and include large amounts of gas and dust.

- <u>Elliptical galaxies</u> tend to have football-like shapes, include stars in random orbits, and have very little interstellar gas or dust.

- <u>Spiral galaxies</u> are shaped like rotating disks with bulging centers where stars are concentrated, and the outer parts of the galaxy rotate at the same speed as the inner parts. The stars are not orbiting as individuals but are rotating as if on a gigantic wheel.

It appears that the shape of galaxies is age related. Young galaxies start out as clusters of stars, gas and dust that aggregate together without discernable shape. Gradually, more stars are formed from the gas and dust, orbiting solar and binary star systems develop, and the collective gravitational harmony gives the galaxy a 3-dimensional elliptical shape. When a huge star in an elliptical galaxy collapses, a super massive black hole is created that draws stars together into a tighter formation that spirals around this black hole, which then becomes the center of that galaxy. It is the rotational energy of the black hole that give this kind of galaxy its disk-like shape.

The pattern is that stars congregate to form irregular galaxies, which eventually evolve into elliptical galaxies, some of which become spiral galaxies. Although the spiral is the latest stage of development, some of the oldest galaxies are the elliptical ones that have not yet given birth to a black hole.

Black Holes

Donald Lynden-Bell is an astrophysicist best known for his theories that galaxies contain super massive black holes at their centers, and that each of these black holes is surrounded by a quasar, a compact region that gives off intense radiation. He estimates that the mass-equivalent of energy required to give the emission of a quasar is 10 million times the mass of our sun.

The accepted explanation for a black hole is that when an immense star exhausts its nuclear fuel, it collapses under its own gravity and becomes trapped in a region whose boundary eventually shrinks to zero size. All the matter in the star becomes compressed into an infinitesimal point with intense gravitational attraction that sucks everything nearby into it. Hubble observations reveal disk-shaped clouds of dust and gases surrounding black holes, probably in the process of being sucked into it.

Billions of stars that are far enough away that they cannot be sucked in, instead spiral around the black hole. It is the spinning motion of the galaxy which gives it its characteristic disk-like shape. Note that these billions of stars are spiraling, not orbiting, around the black hole. They retain their relative positions with respect to each other – rather like specks of dust on a record turntable. Black holes are thus the driving machines responsible for the size and shapes of spiral galaxies.

The gravitational attraction of the black hole from is immeasurably stronger than the star had before it collapsed. So how or from where does a collapsing star acquire its phenomenal increase in its energy of attraction? It can only come from nonphysical reality.

Black holes with zero mass have gravitational attraction powerful enough to gather billions of stars and shape them into galaxies. This is a phenomenon that physics cannot explain. The equation for calculating the force of attraction (F) between two objects is $F = GMm/R^2$, where M and m are the masses of the two objects in question. If the mass of the black hole (M) equals zero, then F also equals zero.

To have zero mass means having no physical existence at all. Whatever intense energy of attraction is coming from the black hole thus must be nonphysical in nature – leaving us with the inescapable conclusion that gravity must ultimately have a nonphysical cause.

When a star collapses and disappears, it becomes part of a chain reaction, spewing out energy and particles from its black hole many thousands of light-years away, eventually giving birth to new stars, some of which die to create new black holes ... and so on.

In 1974, Stephen Hawking provided a theoretical argument for the existence of electromagnetic radiation released by black holes, due to the quantum effects observed near the gravitational boundary surrounding the black hole. Matter is drawn into the black hole and energy is propelled out of it (as radiation and particles). This form of radiation may come from the result of pre-physical particles being boosted by the black hole's gravitation into becoming real particles.

Hubble telescopic observations reveal jets of energy beaming out from two sides of black holes, at right angles to their spin. These powerful jets can travel huge distances, hundreds of thousands of light-years away from their source.

- In 2012, an X-ray of a distant black hole of some 12.4 billion light-years away showed it emitting an enormous stream of radiation about 200,000 light-years across – or twice the breadth of the Milky Way.

- In 2016, astronomers observed bursts of visible light being released from a black hole in the Cygnus constellation, about 7,800 light-years from Earth.

- In 2016, a huge jet stream of energy and particles extending over 300,000 light-years was observed being ejected from the center of the Pictor A galaxy, 485 million light-years from Earth.

Hawking states, *"As a black hole gives off particles and radiation, it will lose mass. This will cause black holes to get smaller and to send out particles more rapidly. Eventually it will get down to zero mass and will disappear completely."* The idea is that the immense energy released from a black hole could create a little baby "universe" of its own.

Hawking continues, *"A small self-contained universe branches off from our region of the universe ... This baby universe may join on again to our region of spacetime. If it does, it would appear to be another black hole that formed and then evaporated. Particles that fell into one black hole would appear as particles emitted by another black hole, and vice versa."*

Stephen Hawking's theory about black holes spawning baby universes is thus an argument for continuous creation – creation that includes periodic explosions. He says, *"In a black hole, matter collapses and is lost forever, but new matter is created in its place."* The final stage of evaporation of a black hole ends in a tremendous explosion, a mini version of the hypothetical big bang.

Stephen Hawking has said, *"The universe may not have any beginning or end"*. The universe has no boundary, neither in time nor space. The universe simply is.

Some big bang proponents accept the part that black holes play in a recycling or re-creative process but stubbornly hang on to the singularity idea. They imagine that the whole process was set in motion by one primordial black hole that was the ultimate cause of everything – leaving unanswered the questions of how, when, and where was the gigantic star that collapsed to form the first black hole – or how was the primeval black hole created if not from the collapse of a super massive star.

A Creation Field

Black hole cosmic recycling explains how matter from old stars is transformed into matter that becomes new stars. If that was all that was going on, then the total matter in the universe would be constant. This is not what satellite images reveal, however. Ancient galaxies viewed from extreme distances are much smaller and less developed than mature galaxies viewed at closer range. Somehow, increasing numbers of stars are being newly created through time. The universe may not be expanding in space, but it surely appears to be increasing in total mass.

In his steady state theory of the infinite universe, Fred Hoyle postulated that new matter is continually created through something called the "creation field". In physics, a field is a region in which a force is effective. In this case, the entire universe is the creation field. Nonphysical reality is continually creating new matter everywhere, as particles that successively become atoms, molecules, gases, and then stars.

Some scientists theorize that in outer space there may be particles popping in and out of existence, just as electrons pop in and out of existence in subatomic space. Speaking of subatomic space, if the atom is continually being re-created at the quantum level, then so is the entire universe – atom by atom.

Cosmic Web

In 2014, a cosmic web of diffuse gas that connects galaxies throughout the universe became visible – thanks to a quasar, 10 billion light-years away, that served as a cosmic spotlight. Computer simulations suggest that the long filaments in the web contain ionized hydrogen gas.

Galaxies appear to sit within these gaseous filaments that stretch throughout the universe, rather like acupuncture meridians throughout the human body. Perhaps these filaments are part of the creation field wherein particles become hydrogen, then helium, then new stars.

The cosmic web may be how nonphysical reality creates new galaxies, either continuously or at intervals, by injecting pure energy into the system that becomes particles, then atoms and then hydrogen gas that is distributed wherever it needs to go throughout a network of filaments.

Plasma Cosmology

Over 99 percent of the matter in the universe appears to be in the form of plasma (hot electrically conductive gases). In plasma, electrons are stripped off by intense heat, allowing them to flow freely. Space is alive with vast networks of electrical currents and powerful magnetic fields that have the potential to create matter.

Big bang proponents see the universe only in terms of gravity. They ignore the important role that electromagnetism (EM) plays. Of the two forces, EM energy may be the primary one. Until matter has actually formed, gravity is of no consequence.

Endless Possibilities

If the universe has no boundaries in time or space, then it is without limits of any kind. It is a universe of endless possibilities.

Some theorists speculate that there may be multiple universes. This is a contradiction in terms. If we define "universe" to include everything which exists, then this term necessarily includes not only the known universe but all aspects which are yet unknown. If something other than the known universe should one day be discovered, by its very existence it is and always has been part of the one and only entire universe.

Other theorists speculate that there may be parallel universes – one or more cosmos just like the one we know, in some of which parallels we may be living multiple lives. If subatomic particles and atoms can manifest in more than one place at a time, then it is possible that our entire bodies could also manifest in more than one place. Again, if these other places exist, they are part of the one and only universe.

In 2016, surveys taken by NASA's Hubble Space Telescope suggest that there may be as many as 2 trillion galaxies in the visible universe. Other evidence suggests that there may be 200 billion stars in the Milky Way galaxy alone. Assuming our galaxy is of typical size, there could be 200 billion X 2 trillion = 400 trillion stars in the known universe, many or most of which could be hosting complete planetary systems.

In 2015, NASA's Keplar telescope identified 4,706 planets in regions around stars in the Milky Way where liquid water could exist on the surface of rocky planets in habitable temperature zones. Statistical analysis predicts (with a confidence level of 99%) that 1,284 of these planets are capable of sustaining life – and that a further 1,327 are likely capable of sustaining human life.

For purposes of projection, let's assume there are 2,000 habitable planets in our galaxy. Two thousand X 2 trillion galaxies = 4 quadrillion habitable planets in the known universe.

The estimate of four quadrillion habitable planets assumes that the known universe is all there is. If the universe is infinite, then there may be an infinite number of planets capable of hosting life, including human life.

Time is a Thought

Time is in the universe; the universe is not in time

Time is a thought – a mental abstraction we created to measure duration, rates of change, and successions of events in our physical world. We have conveniently linked our units of time measurement to cycles of nature – days/hours/minutes/seconds to the rotation of Earth on its axis – years/months/weeks to Earth's orbit around the sun.

In medieval times, peasants divided the time between sunrise and sunset into 12 hours – thus, hours were longer in summer and shorter in winter. Now that we have standardized the hour, every day in every locality has its own timing (e.g., noon varies by location). Time can only be specific to a specific place. Einstein expressed this as a relativistic principle: time travels at a different speed depending on where you are. This applies both to where you are on the globe and how high you are above sea level. Time travels faster at higher altitudes because of the further distance from the center of the Earth's rotation.

Time measures the motion of an object relative to a specific location. If there are no physical objects, then time has no meaning. Time cannot exist independently of matter.

From our physical perspective, it could appear that matter has always existed – but "always" is a subtle way of smuggling in the assumption that the universe had a beginning. From the timeless perspective of nonphysical reality, there is no such thing as a beginning. Everything simply is.

Time has no beginning. There is no such thing as a preceding point in time when time began.

Time has no ending. There is no such thing as a point in time at which time will end.

Time is relative. There is no such thing as absolute time.

Time is a measure of duration, e.g., how long it takes an object to move from point A to point B – which depends on (i) the distance between A and B, and (ii) the velocity with which the object is moving. Time only has meaning relative to the initial reference point, A.

> **Thought experiment**: imagine yourself looking in on a three-dimensional world that consists only of fixed objects. Nothing moves. Whenever you come back again, no matter how much time may have elapsed in your world, nothing will have changed in this stationary world. It is and will always be exactly as it has always been. Time has no meaning in this motionless world.
>
> **Thought experiment**: imagine yourself looking into a chamber in which a vacuum has been created. There is nothing in this chamber, not a single molecule nor even one particle of anything. No matter how many times you come back, the chamber always looks the same. Time exists for you who is moving about, but not in the vacuum state in which there is nothing capable of motion.

Beyond Time

Time is a measurement that depends on distance and velocity. Mathematically: $time(t) = distance(d) \div velocity(v)$. If distance equals zero, then time also equals zero. If there are no objects located at various physical locations, then the concept of distance has no meaning and neither does the concept of time.

$$t = \frac{d}{v} \quad \text{If d = 0, then } t = \frac{0}{v} = 0.$$

If distance is zero, then time is zero or nonexistent.

$$v = \frac{d}{t} \quad \text{If t = 0, then } v = \frac{d}{0} = \infty.$$

If time is zero, then velocity is infinite or instant.

The latter equation tells us that if it takes zero time to travel from A to B, then whatever is happening must be nonphysical – because in our physical world, nothing travels faster than light. In quantum mechanics, sometimes a change to one particle instantly creates the identical change in a distant particle. Since no time has elapsed between the simultaneously observed changes, both particles must have been affected by a nonphysical force that is instantaneous. That force can only be the power of thought.

Faster than Light

The speed of light is 186,000 miles per second. This is the limiting velocity in the physical universe. Only massless particles/waves of light (and radiation) can travel at this speed. Anything with mass cannot. According to Einstein's energy equivalence equation, $E = mc^2$, It would take an infinite amount of energy to propel even one gram of matter at light speed – clearly an impossibility.

Thought is instant because it transcends time and space. Thought has no physical destination to reach, no physical space through which to pass, and therefore takes no time to get wherever it is going. This makes thought the fastest speed in the universe, infinitely faster than light.

Thoughts are abstract, intangible, and without any physical properties. They exist only in nonphysical reality and yet are how we humans create whatever we wish in our physical world. Every invention ever created began as a thought in someone's mind.

As physical beings, we can only infer the existence of nonphysical reality. Because the nonphysical transcends time and space, we can also deduce that whatever happens in that realm must happen instantly – at the speed of thought.

The ultimate limiting velocity in physical reality is the speed of light. Whenever we observe a phenomenon in quantum mechanics happening faster than light (e.g., at the speed of thought), then we know that whatever is being observed must have a nonphysical cause.

Spacetime, an Illusion

Spacetime is an abstract mathematical model that fuses the three dimensions of physical space and the nonphysical dimension of time into a single 4-dimensional continuum. This is an interesting excursion that bears no relation to reality.

Suppose a world of 2-dimensions could exist, and you wish to represent that world on a 3-dimensional graph. How would you know if that circle you see is a sphere, a cone, a dome, a cylinder, or something else? It is not possible to extrapolate meaningful information from two dimensions into three, nor from three into four.

General relativity is a theory of gravitation developed by Einstein between 1907 and 1915, in which he predicted that gravity bends light. Experiments in 1919 and 1922 during total solar eclipses appeared to confirm that this is so. Einstein used this 4-dimensional continuum model to illustrate graphically his theory of how the mass of an object supposedly curves spacetime. But spacetime cannot be bent. Spacetime is not real. It is an illusion.

All that Einstein accomplished with spacetime modelling was a visual depiction of his theory. This model is incapable of explaining cause-and-effect.

Time is an abstract measurement of the duration and sequences of events and the changing positions of objects within 3-dimensional physical space. It is impossible (and nonsensical) to extract nonphysical time from physical space and project it onto a separate axis with its own independent set of reference points describing something that supposedly exists independently of the 3-D.

The three dimensions of space are real and tangible. The dimension of time is abstract and intangible. Whatever model you create that includes mathematical measurements of an intangible dimension (i.e., a dimension without physicality) cannot possibly be real. To believe in spacetime is to believe in at least one direction to which one cannot point.

Conscious Time

In your thoughts, time has little or no meaning. When you remember something dramatic that happened, mentally you feel as if you are right back there in that moment, either observing or re-experiencing it. Whenever you re-encounter a good friend whom you haven't seen for many years, although your bodies have aged, emotionally it feels that no time at all has elapsed. From the perspective of your nonphysical mind, time has no meaning.

In your dreams, you don't think of yourself as being any specific age. Awakening in the morning, before any daily thoughts creep in, feels the same to you now as it has always felt during your entire life.

You can also recall personal events from your distant past as clearly as if they happened yesterday. Your mind is nonphysical and timeless.

You are the consciousness that occupies your physical body. Your body is subject to time; but you, its occupant, are not. The real you, who you truly are, is timeless and ageless. You are a nonphysical being who experiences through your physical body.

How you experience time in any moment is a function of where you focus your thoughts. If you are engrossed in an exciting project, you may feel that only an hour has elapsed – and be amazed when the clock indicates you have been at it for three hours.

A familiar train ride can seem to take forever or pass very quickly, depending on where you are putting your attention. If your mind is clear and you are looking at scenery as if for the first time, you will have a very different experience of how long that trip took than if you are preoccupied with stressful or worrisome thoughts.

Have you ever taken a familiar short trip and arrived at a time that doesn't seem possible? Maybe you had left late and were focusing all your attention on getting to an appointment on time. Your mind was rushing, but traffic was not. It was moving at the same pace as always. But when you arrived, your watch showed that only 22 minutes had elapsed rather than the usual 30. Perhaps our conscious intentions can influence time.

Endless Space

Space is in the universe; the universe is not in space.

Just as the universe had no beginning in time, so also did it not have any beginning in space. The universe did not begin anywhere. It isn't located anywhere.

Space is relative. There is no absolute space.

Space is a measure of the position of objects relative to each other. If there are no physical objects, then space has no meaning. Space cannot exist independently of matter.

When we view the known universe, we see stars and galaxies separated from each other by vast expanses of space. When we peek inside the atom, we see particles separated from each other by vast amounts of space. In both cases, space has meaning only in relation to the objects within it. Without objects, the concept of space has no meaning.

The universe has no boundaries. A boundary is a line dividing an area of something from an area of something else. The universe is defined as everything which exists. To imagine a boundary is to believe that something exists on the other side of existence, a self-evident contradiction.

The universe has no outer edge. To assume that it does is to believe that beyond this edge is empty space. Without physical objects on the other side of such edge, the concept of space is meaningless. So is the concept of "empty", which implies that there are physical boundaries (i.e., external to the universe) inside of which there is a total absence of any kind of matter. Matter, space, edges, and boundaries exist only within the universe and not without.

There is a theory that the supposed big bang may be one of perhaps an infinite series of big bangs, that each big bang creates its own virtual bubble, and the universe we know is simply one of those bubbles. A bubble in this context is an imagined boundary beyond which some form of space is presumed to exist.

Neither the entire universe nor any sub-part of it exists anywhere in space. Big bang explosions cannot exist anywhere as separate, disconnected bubbles or boundaries within space – because neither space nor supposed partitions within space can exist independently of existence itself. The one universe includes everything that is.

Relative Location

Because space is shapeless and limitless, there is no such thing as absolute location. There is only where we are (in the Milky Way) relative to the positions of other galaxies.

We cannot tell how far we are from the center of the universe, because it has no center. There is also no absolute direction within the universe – no up, down, sideways – no north, south, east or west. All we can say is that galaxy X is so many light-years away.

When scientists took measurements suggesting that galaxies are accelerating away from the Milky Way, they presumed that only a super colossal explosion (big bang) could have caused this to happen. If so, where could this explosion have possibly been located?

If an explosion had created the universe, galaxies would be travelling on radial paths away from what would have become the center of the universe. However, there is no evidence of galaxies moving on radial paths -- and how could there be? The universe has no center.

Presumption of a causal explosion leads to another presumption, that the universe is spheroid in shape – that all of existence is contained within an unfathomably large sphere. How can existence be restricted within any shape without presupposing that something already exists outside the boundary of that shape?

When our space telescopes pick up images from 13.4 billion light-years distance, the universe looks structurally the same in every direction. It makes us feel as if we are at the center of a giant sphere, exploring its perimeter. This is the same feeling we would get if cast adrift in a lifeboat with no land in sight. The horizon would look the same in every direction, to the limits of our field of vision. We have no way of knowing what lies beyond that horizon.

What our space telescopes pick up at extreme distances is a spherical horizon, a view from a cone whose vertex is at the observer's eye. The radius of our spherical horizon is 13.4 billion light-years. Its diameter is 26.8 billion light-years. And we have no way of knowing what lies beyond that horizon.

Galaxy GN-z11 is 13.3 billion light-years away from us. Suppose there is an advanced civilization there with technology comparable to our own. Their spherical horizon would similarly have a radius of 13.4 billion light-years and a diameter of 26.8 billion light years.

The above two spherical horizons intersect and overlap by 0.2 billion light years. Therefore, the combined breadth of the two horizons is 53.4 billion light-years (a measurement that is in obvious conflict with the big bang assumption that the universe began 13.8 billion years ago).

Spatial horizons tell us absolutely nothing about (a) the size or shape of the universe, (b) how old it is, or (c) how it was created. Collective spatial horizons from every known galaxy suggest only one thing: the universe is infinite.

Looking Back in Time

***We are unable to see beyond the spherical horizon
created by our space telescope.***

Popular science shows on television tell us that the Hubble telescope is looking back in time. This is not strictly accurate. Space telescopes are in the present moment, picking up electromagnetic radiation from far distant galaxies.

The language is confusing. A light-year is a measure of distance, not time. But because we associate "year" with time, every time our brain hears "light-year" it automatically thinks time.

We could clear this confusion by renaming our standard measure so that it is associated only with distance. Just for fun, let's create the term "bleem" – a neutral word that has no association with time – and has only the meaning we give it, such as the following:

    I bleem = 6 trillion miles
           = 9.6 trillion kilometers
           = 1 Earth light-year
           = 0.53 Mars light-year

Now, when we say that the Orion nebula is 1,344 bleems away, everyone knows we are talking about distance.

    Thought experiment: Imagine you are looking in on the universe from another dimension that is beyond time. You happen to observe a super colossal explosion in one galaxy – and in the same instant view another super colossal explosion in another galaxy that is 10 billion bleems away from the other. You know for a fact that the two explosions happened simultaneously.

Observers in galaxy A, however, won't have any evidence of the explosion in galaxy B until the light from it reaches them 10 billion years after the event – and vice versa.

Observers in galaxy A will have evidence of their own explosion relatively soon after the event, because of the much shorter distance light is travelling. From their perspective, explosion A was in recent history and billions of years will have passed before they become aware of the explosion in B – and thus they will have no way of knowing that it was simultaneous with their own. (Vice versa for observers in B).

This thought experiment demonstrates:
1. There is no absolute time.
2. Time is relative to the location from which we are experiencing.
3. If a galaxy is at an extreme distance, we have no viable way to compare its history to our own.

Recently discovered Galaxy GN-z11 is 13.3 billion bleems away from the Milky Way. This we can tell from the image our space telescope received from the photons that were emitted from it 13.3 billion years ago. This image is a snapshot of a moment in time in the ancient history of GN-z11 – and tells us nothing about our history. It also tells us nothing about GN-z11's present, how large it may be today, or even if it is still located in the same position relative to Earth.

If the movement of galaxies is an illusion, then GN-z11 is in the same position it has always been – 13.3 billion bleems away from the Milky Way. GN-z11 now presents this conundrum for big bang theorists:

1. If galaxies are in motion, then GN-z11 and the Milky Way would have been 13.3 bleems apart 13.3 billion years ago. If the "big bang" happened 13.8 billion years ago, then GN-z11 grew from nothing to be a full-sized galaxy that in 500 million years had travelled 127 septillion kilometers (13.3 billion bleems) – at a velocity that would have had to have been over 82,000 times the speed of light. This is beyond ridiculous.

2. If galaxies are in motion, then GNz-11 and the Milky Way have been continuing to travel further apart for the last 13.3 billion years since our current snapshot of their distance of 13.3 bleems. In other words, these galaxies have been in motion for at least 26.6 billion years – predating the alleged "big bang" by 12.8 billion years.

3. If galaxies are not in motion, then there was never any need to postulate a big bang.

Faster than Light

Thought is faster than light.

The speed of light is 186,000 miles per second (299,800 km./sec.), the maximum velocity of anything in the physical universe. Only photons can travel at light speed, however. Anything that has mass would require incredible energy to get it to approach light speed, if it is even possible.

The reason photons can travel so fast is that they have zero mass. Photons are micro packets of pure energy given off by atoms when electrons drop from higher energy orbitals to lower energy orbitals. Light speed is thus the maximum velocity that **energy** can travel in the universe and has little or nothing to do with how fast matter can travel.

If photons had mass, light would gradually slow down as it entered gravitational fields of galaxies, stars, and planets. There would be no way to make sense out of the visual distortions. Cosmology would be unknowable.

We measure the distance between galaxies in terms of light-years (the distance light travels in an average solar year) or approximately 5.88 trillion miles. Our galaxy is about 100,000 light-years across. Thus, it would take a space ship travelling at the speed of light 100,000 years to travel from one side to the other. This is a theoretical consideration that could never actually happen. The amount of energy required to accelerate an object increases exponentially as its velocity increases. To attain light speed would require an infinite amount of energy – clearly an impossibility. We will never be able to propel spacecraft at anywhere near the speed of light.

Particles or information travelling at (or above) the speed of light is beyond anything physics can explain – and, yet, it appears to happen. Metaphysics, however, has explanations for such phenomena.

Speed of Thought

The speed of thought is instantaneous, infinitely faster than the speed of light. Thought is a form of nonphysical energy that transcends the limits of time and space.

We see evidence of thought energy in quantum physics in "entanglement", in which pairs of particles are generated whereby the state of one particle instantly affects the state of the other, even though these particles may be separated by large distances. If you manipulate what is going on in one location, you automatically change what is going on in the other location, even when there is no physical force connecting the two and no time for information to pass between them.

The time between the detection of one particle and its twin is less than the time it takes for light to travel from one detection apparatus to the other. Therefore, whatever communication is occurring between the particles must be happening faster than light speed – i.e., at the speed of thought.

The logic is thus: if two objects are in fact connected and there is no physical force between these objects, then whatever is connecting them must be nonphysical. Perhaps the two particles are manifestations of the same thought. Although lacking in consciousness themselves, the particles are apparently responding to mental energy from a conscious source.

Everything is Energy

Matter can be converted to energy, and vice versa

Matter and energy are two different manifestations of the same energy of which everything is composed. Everything in the universe is energy – including planets, inanimate objects, plants, animals, and ourselves. Life is energy. Thoughts are energy.

Matter is energy compressed into atoms. It is dense, concentrated energy vibrating at a frequency that makes it appear solid to our physical senses. The entire material world is made only of energy (as light) that flashes on and off.

Every substance vibrates to a specific frequency that gives it its unique characteristics and appearance. In the laboratory, spectrophotometry measures the frequencies emanating from each substance to determine the elements of which it is composed. Through telescopes, scientists can tell the compositions of various stars by the colors of light they give off.

It is activity at the subatomic level which creates the unique frequency for each molecule. Electrons within an atom occupy energy shells or spatial domains known as orbitals. Each orbital has certain energetic frequency characteristics, depending on the type and molecular weight of the atom. When an electron spontaneously jumps to a higher orbital, it absorbs light of a specific frequency. When it spontaneously drops to a lower orbital, the electron gives off light of that same frequency. This is orchestration at its finest: every element and every substance in the universe has its own unique energetic signature.

Each element emits its unique frequency (color) of light, thus enabling scientists to determine the composition of distant stars. If the color spectrum of elements was not part of a deliberate plan, then it is certainly a most useful side benefit.

Energy in various forms is around us, within us, and powers our world. We receive light and heat from the sun. We hear voices, music, nature and noise by means of sound waves. Our bodies receive caloric energy from food and physiological energy from the internal breakdown of ATP (adenosine triphosphate). Plants receive energy from the sun via photosynthesis. We witness the kinetic energy of vehicles in motion. We listen to radio and watch television by means of transmitted electromagnetic waves. Our homes, computers and electronic devices are powered by electricity.

All the above forms of energy have physical causes. Light is given off by glowing objects due to thermal motion of atoms within them. Heat arises from the motion of molecules and may be transferred by conduction, convection, or radiation. Electromagnetism is a force that arises between particles with electric charges. Electricity results from the existence of charged particles (usually electrons) either statically as an accumulation of charge, or dynamically as a current.

The above forms of energy also have physical properties -- such as vibration, oscillation, wavelength, amplitude, frequency, intensity, recurrence – all of which can be measured by scientific instrumentation. Our experience of energy is that it has a physical cause, that it is converted from matter, either directly (e.g., combustion, explosion) or indirectly by activity within that matter (e.g., radiation, conduction). It is only natural to assume that whenever we encounter energy in any form that it must have a physical cause, but not necessarily so.

Matter and energy are known to be interconvertible, by Albert Einstein's famous $E = mc^2$ equation. This means that (a) matter can be converted into energy, and (b) energy can be converted into matter. Although scientists have been unable to create matter from energy in laboratories, this may be happening continuously in the real world, in subtle ways that elude our senses and our instrumentation.

The assumption that all energy has a physical cause is unsupported at the subatomic level. Quantum physicists observe intermittent and elusive energetic phenomena, and to explain these anomalies postulate the existence of undetectable (i.e., virtual, hypothetical) particles or waves which must have caused them. This is a case of creating evidence to support a theory rather than modifying the theory to explain the known evidence. Not all energy has a physical cause. Some energetic phenomena, including events in the cosmos and within atoms, may be created in whole or in part by nonphysical energy.

Gravitational Energy

Gravitational waves transport energy in the form of gravitational radiation. Their existence was predicted by Albert Einstein, in 1916.

Gravitational waves were first detected coming from outer space in 2016 – from the merger of two black holes that happened about 1.3 billion light-years ago. The energy released by this merger was equivalent to 60 times the mass of our sun (or 20 million times the mass of Earth), making this the most powerful energetic event ever witnessed by humans

The energy transported by gravitational waves is very different from electromagnetic energy, the latter of which is transmitted by synchronized vibrations that are incapable of being transmitted through empty space. It took very sensitive instrumentation to detect gravitational waves emitted from space by the mass equivalent of 20 million earth-sized planets. The implication is that the gravitational waves emitted by a body as small as Earth are so weak as to be undetectable by any known means.

Every bit of matter in the universe attracts every other bit of matter, and gravitational waves may be how this happens. Knowing how gravity works, however, does not tell us why there is gravity. Gravitational attraction is a function only of mass and distance and has absolutely nothing to do with physical properties. No matter how dissimilar two bodies may be in molecular configuration or chemical composition, they always attract each other. If gravity does not depend on physicality, then ultimately it must be nonphysical in origin.

An Illusion

Einstein once said, *"Everything in life is vibration"*. Atoms which to us appear solid are vortices of energy that are constantly spinning and vibrating. It is only because matter vibrates within a specific frequency range that we can perceive it with our physical senses.

That which we call solid is so only in our experience but not in ultimate reality. Our senses of sight and touch are how our bodies (composed of matter) experience the physical aspects of other forms of matter. These senses are incapable of detecting (a) the energetic or vibratory aspects of matter, and (b) forms of energy that vibrate at higher frequencies than that of matter. Sight and touch thus create the illusion that only solid objects are real.

Our sense of hearing makes us aware of sound vibration within the frequency range of 20 Hz to 20,000 Hz. It cannot detect ultrasound frequencies (above 20,000 Hz) nor the infrasound waves (below 20Hz) that developing earthquakes give off. This sense thus creates the illusion that the only sound which exists is that which we can hear.

Specialized instrumentation can measure such forms of energy as radio waves, television waves, electromagnetic radiation, microwaves, ionizing radiation, seismic vibrations, infrared energy, ultraviolet energy, and kinetic energy. The illusion created by this instrumentation is that the only energy that exists is that which can be detected.

Conservation of Energy

The conservation of energy principle states that energy can neither be created nor destroyed but only changed from one form to another, or transferred from one object to another. We experience matter being transformed into various forms of energy (e.g., combustion propels our cars, fires cook and heat, food energizes our bodies, radio-active decay facilitates medical diagnoses). We also experience one form of energy being converted into another (e.g., sunlight into photosynthesis, electricity into artificial lighting, radio waves into sound, microwaves into heat). We have no direct experience of energy being converted into matter, yet this may be how all matter in the universe ultimately came to be.

Einstein's equation, $E = mc^2$, states the energy equivalence of any given amount of mass, or how much energy a unit of matter could produce if all of it were converted to energy. Transposing this same equation to read $m = E \div c^2$ gives us the formula for calculating the mass equivalent of energy, or how much matter can potentially be produced from a given amount of energy.

There are three principles that we can infer from observing how energy works:

(a) **Energy is purposeful**. Every form of energy works in a specific way to produce a specific outcome.

(b) **Energy is never wasted**. Energy cannot be destroyed. It can only be converted to other forms of energy, including matter.

(c) **Energy follows the path of least resistance**. Energy travels in a straight line until it bumps into or is deflected by something, or until the medium it is travelling in changes.

Nonphysical energy appears to follow these same principles. The nonphysical (a) interacts with the physical for specific purpose, (b) does only what physical energy cannot, and (c) uses the most direct means possible.

Nonphysical Energy

If there is a nonphysical reality which creates and supports everything in the universe, of what does this nonphysical reality consist? The most plausible answer is that it is pure primary energy – pure in that it has no physical attributes – primary in that it is the ultimate first cause of everything.

There is no language to describe what nonphysical reality may be like. The closest we can come is to say what it is not. It is completely without physicality of any kind – which means no objects, no dimensions, no motion, no time, no space, and no limits. We cannot call it an energy field because "field" implies the existence of physical boundaries. Nor can we call it a "void" because this term implies a physical space with nothing in it. This primary energy which created everything else is timeless, dimensionless, and limitless (i.e., infinite).

Life Energy

If you plant a seed in your garden and conditions are right, it grows. If you boil the seed first, it cannot grow. If you grind the raw seed to a powder, the powder cannot grow. In both cases, while not altering the chemical composition of the seed, you have made it an unfavorable host for life energy.

Life energy is real and has no physical cause. It isn't created by chemical reactions or electromagnetism, nor does evolutionary theory have any explanation for how inert substances could possibly have become living organisms. The same nonphysical energy which created the universe also created life, when the timing was right and on planets where conditions are favorable.

Meridians of Energy

In Chinese philosophy, ch'i is a form of life energy that flows from the environment into the body via portals of entry (acupoints) on the skin, which are inlets into a specialized meridian system that distribute this energy to organs and tissues. This meridian system is a physical-nonphysical interface.

Meridians can be viewed as electrical circuits which connect superficial entry points on the skin to deeper organ structures, as verified by measurements of electrical skin resistance in and around the acupoints. Meridians also distribute a specialized type of electrolytic fluid which communicates certain types of energetic information to various tissues in the body.

In the 1960s, Kim Bong Han experimented with the acupuncture meridians of animals. He injected radioactive P^{32} (an isotope of phosphorus) into rabbit acupoints and followed the uptake of this substance into surrounding tissues. Using microautoradiography, he discovered that the P^{32} was actively taken up along a fine duct-like system of tubules (averaging 1.0 micron in diameter) which followed the classical path of acupuncture meridians. Kim also found special small corpuscles in these ducts which lie beneath and correspond with the classical acupuncture points on the meridians. Pierre de Vernejoul and Jeane-Claude Darras have confirmed Kim's findings in humans. In 2016, scientists at Seoul National University confirmed the existence of meridians in humans, which they refer to as the "primo-vascular system".

Kim found that within the embryonic chick, meridian ducts were formed within 15 hours of conception. Formation of these ducts precedes both the development of organs and the vascular and lymphatic systems. As blood vessels develop, they grow around the meridians.

Based on many experiments, Kim concluded that the meridian system interconnects with all cell nuclei of the tissues. Some type of information flows through the meridians to the DNA control centers of the cells, providing an energetic guidance system to the developing cells of the body. The meridian system may thus play a part in both the differentiation (specialization) and replication of all cells in the body.

Fluid extracted from meridian tubules reveals high concentrations of nucleic acids (DNA, RNA), amino acids, and hormones (adrenalin, corticosteroids, estrogen). Twice the blood level of adrenalin has been found in meridians and over 10 times the blood level of adrenalin has been found in acupoints.

The presence of nucleic acids and hormones such as corticosteroids and estrogen strongly suggest that the meridian system may influence endocrine balance in humans.

Imbalances in the acupuncture meridians precede and predict physical organ dysfunctions. This relationship strongly suggests that energy imbalances may be a causative factor in the development of diseases and abnormal physical states.

Orgone

Orgone may be a form of universal life energy occurring subtly in all of nature. During the 1930s and 1940s, Wilhelm Reich built specialized energy capacitors for accumulating and concentrating ambient orgone (life-force) energy from the atmosphere. Reich and other medical researchers reported successes in treating cancer with these devices, which were essentially insulated Faraday cages that used a type of Geiger counter as a measuring device. In 1986, scientists at the University of Marburg, Germany, published the results of a blind study which showed that 30-minute orgone accumulator treatments caused consistent, positive psycho-physiological effects not seen with the inert machine used as a control.

In several experiments, Reich placed protozoa inside an orgone accumulator. He observed flashes of bluish light and bluish vapors originating from the cultures. When the culture dishes were taken out, this visible radiation could still be seen by the naked eye. What is especially noteworthy is that small, round blue-glowing vesicles had formed along the edges of the culture and swam off as new protozoa. Reich thus demonstrated that orgone energy can speed up the replication of unicellular organisms.

Orgone may be what Indian philosophy calls "prana", the life energy we breathe in from the atmosphere along with the air that we inhale. In addition to vitalizing bodily cells, orgone/prana may also provide a fuel source to the body in addition to its caloric intake from food. Mystics who can meditate for prolonged periods without food attribute their ability to do so to prana, the vital force in Hinduism.

Survival guides tell us that the human body can last from 30 to 40 days without food. However, there have been some cases of people lost in the wilderness or trapped under avalanches who have survived from 46 to 60 days without food. There have also been some prisoners on hunger strikes who lasted from 53 to 75 days. Perhaps orgone can take over where nutrition leaves off.

Prahlad Jani (*1929* -), an Indian breatharian ascetic, claims to have lived without food and water since 1940, which claim his devotees confirm. Recent photos of Jani reveal his body to be emaciated and with atrophied jaw muscles. He is, however, active and very much alive.

In 2003, Dr. Sudhir Shah and other physicians at Sterling Hospitals in Gujarat kept Jani in a sealed room for 10 days and observed that in addition to neither eating nor drinking during that time, he also passed no urine or stool.

In 2010, Jani was again tested at Sterling Hospitals by Sudhir Shah and a team of 35 researchers from the Indian Defence Institute of Physiology and Allied Sciences (DIPAS) and other organizations. Jani was given round the clock surveillance, including closed circuit television and personal observation. After 15 days of neither eating, drinking, nor going to the toilet, all medical tests performed on Jani were reported as normal and researchers described him as being in better health than someone half his age. The director of DIPAS said the results of the observations could "tremendously benefit mankind", as well as "soldiers, victims of calamities, and astronauts", all of whom may have to survive without food or water for long durations.

Under favorable conditions, it is possible to see orgone energy in the atmosphere. In natural surroundings on warm days with a blue sky, relax, look up at the sky and defocus your vision or look from the corner of your eye. You may see tiny orbs of what look like dancing particles of light. If your eyes are particularly sensitive, you may also see orgone accumulating around the periphery of trees.

Pure life energy is nonphysical and invisible. It only becomes detectable (e.g., as orgone) after it differentiates itself for purpose. Because ch'i and orgone serve different purposes, each may be an individuated manifestation of the same universal life energy.

A Universe of Attraction

Attraction is the underlying principle upon which all of creation, all of science, and all of humanity depends.

The universe and everything in it, including ourselves, could neither exist nor thrive without an orchestration of various forces of attraction. These include:

1. **Gravity**: the physical force of attraction between every particle of matter in the universe and every other particle. Without gravity, there would be no universe – no dust, no stars, no planets, and no life. Everything in the universe exists because of gravity and moves under its influence.

2. **Electromagnetism**: the physical force of attraction between positively and negatively charged particles. This is the force that causes electrons to orbit the nucleus of the atom, just as gravity causes the earth to orbit the sun. Electromagnetism holds atoms together and makes objects appear and feel solid, even though over 99% of the atom is empty space. Magnetic fields assist gravity in the formation of stars and the shaping of galaxies. Earth's magnetic field protects its inhabitants from harmful solar winds.

3. **Nuclear force**: the physical force that holds together protons, neutrons, and the nuclei of atoms. It is alleged to be 137 times stronger than electromagnetism. (Electromagnetism has no effect on neutrons, which are particles without an electric charge.)

Without this strong nuclear force working in harmony with electromagnetism and gravity, the universe could not exist.

4. **Chemical Affinity**: the tendency for certain substances to combine with each other, depending on the compatibility of their molecular structures. This happens by means of covalent bonding (sharing electron pairs between atoms) and ionic bonding (transferring electrons between atoms). Fortunately for us and all other life forms, hydrogen and oxygen have a strong affinity to combine to form water. Without chemical affinity, the universe would probably consist of only gas, dust, and rocks.

5. **Sexual attraction**: the reason why species proliferate. For humans, this attraction is both biological and psychological.

6. **Social Affinity**: we gravitate towards and tend to form relationships with those who share our deepest values and/or common interests.

7. **Mental attraction**: like thoughts attract like thoughts.

    Thought experiment: clear your mind and focus exclusively on one intensely happy thought, perhaps the most exciting thing that ever happened to you or the happiest day in your life. In about 22 seconds, another happy thought will join the first one, as if being attracted magnetically. You can also observe this same phenomenon in people who are chronic complainers: the more they complain, the more they attract other things about which to complain.

Science can measure how the first four of these forces operate, according to established principles of physics and chemistry. Outcomes of interactions between variables can be predicted with mathematical precision. What science is unable to answer, however, is why these forces exist. The answer is a philosophical one: they have to exist in order to create and maintain the universe. Attraction is the underlying principle upon which all of creation, all of science, and all of humanity depends.

How electromagnetism and chemical affinity work is straightforward. In both cases there is a physical act of completion. Positively and negatively charged particles are energized fragments that seek each other out to complete a stable (uncharged) state. Atoms seek to complete their orbitals with electrons by combining with other atoms.

Gravity is a physical force of attraction without a physical cause. Every bit of matter attracts every other bit of matter for no physical reason: there is no electromagnetism involved, no exchange of particles, no chemical affinity, nothing explicable in scientific terms. It makes no difference how dissimilar in composition or characteristics two objects may be, they attract each other. Physics explains how gravity attracts but has no clue as to why it attracts.

Gravity is a real phenomenon without a physical cause. Therefore, it must have a nonphysical cause. Nonphysical reality appears to imbue every bit of matter with the ability to attract all other matter.

Why there is gravity, the primary force that makes the universe possible, is a mystery to science. Every object in the universe attracts every other object, just because it does – without any physical interaction taking place between them, and regardless of how dissimilar those objects may be. Neither object has anything that it requires from the other; both are complete unto themselves. Gravity is a powerful physical force without a physical explanation. Therefore, the explanation must be nonphysical.

Astronauts in the weightless atmosphere of space have experimented by putting table sugar into closed transparent plastic bags to see what happens.

The sugar molecules are gently attracted to each other by gravity to form clumps that steadily increase in size. This is how planets form: bits of debris steadily attract each other to form increasingly larger conglomerations of matter that eventually become planet sized.

Gravity may also be responsible for cloud formation. Clouds consist of tiny droplets of water or ice crystals that settle on dust particles in the atmosphere and are suspended by upward air pressure created by heat radiating from the ground. The upward air pressure neutralizes the downward pull of Earth's gravity, thus creating the same kind of weightlessness as in the astronauts' sugar experiment. Horizontally acting gravity brings the droplets together to form clouds.

> **Thought experiment**: on a sunny, partly cloudy day, look up at a small cloud in the sky. Focus all your attention on this cloud and tell it (silently) that you wish it to dissipate, that you wish its tiny water droplets to go separate ways rather than cling together. Keep sending the cloud subliminal messages telling it to disappear. Use whatever language makes your intention clear. The cloud will gradually disappear, as if dissolving. Your intention is more powerful than the mild gravitational force that holds the drops of water together. You have just demonstrated that your thoughts can alter physical presentation. Something like this may also happen at the quantum level: your intention may influence how subatomic particles present themselves.

Of all the forces of attraction in the universe, gravity is the primary one. Without gravity, planets would not exist, life forms would never have evolved, and we would not be here now questioning the universe. Gravity is a physical force that has a nonphysical cause. The only plausible explanation is that gravity is a universal energy of attraction that could only have been imprinted into all matter by a universal mind.

A Universe of Light

"Let there be light" may be more than a metaphor.

Light is electromagnetic radiation made visible. When electrons leap from higher to lower orbitals, they give off energy as photons – quanta of pure (massless) energy that we perceive as light. The mystery is why electrons spontaneously jump back and forth between orbitals.

Without light, our atmosphere would never have been oxygenated by photosynthesis. The evolution of all known plant and animal species would never have happened. The only life forms that could exist would be anaerobic bacteria and multicellular anaerobic marine species, such as tiny Loriferans (less than 1 mm.) recently discovered living in sediment under the Mediterranean Sea floor.

Light is the facilitator that makes possible higher life forms (including ourselves). Was it deliberately planned that way, or is it merely a fortunate side benefit of random processes?

Randomness becomes less of a consideration when we examine how electrons make quantum jumps. An electron simply disappears from its orbital, is physically absent for a tiny increment of time (measured in billionths of a second), then suddenly reappears in a different orbital. Why the electron flashes in and out of existence is a question that physics cannot answer – because it has nothing to do with any physical force.

An electron is a tiny quantum of energy that when physically present behaves as an electrically charged particle. For that same packet of energy to reappear after an absence means that it must have had a nonphysical existence while it was not physically present.

The frequency with which an electron makes quantum jumps is the frequency of light emitted by the photons generated during this process. It so happens that every element in the periodic table emits a unique electromagnetic frequency (color) of light, a unique energy "signature". An electron is an electron, regardless of the atom in which it appears – and a photon is a photon, regardless of its source. Since electrons and photons never change, the only factor that accounts for the unique vibration of each element is the specific frequency with which its electrons flash in and out of existence.

That each element has a unique frequency is a boon to science. The various colors of light emitted by stars tell us of which gases they are composed and how old they are. Spectroscopy in the laboratory tells us the elements from which substance are composed – and through carbon dating also tells us how old organic substances are. Were these scientific benefits deliberately planned or simply a side benefit of random processes?

The physical properties of each atom are determined by its unique configuration of protons, neutrons, and electrons – irrespective of how frequently its electrons pop in and out of existence. Quantum jumps impart to each element a unique energetic vibration, which we and our instruments detect within a color spectrum of light. If the electrons within every atom of every substance blinked in and out of existence at the same rate, everything we see in our world would be the same color.

Electrons blinking in and out of existence (and at varying rates) is beyond any physical explanation, and thus must have a nonphysical cause. Therefore, light itself has a nonphysical cause. It thus appears that a universal mind decided not only that there should be light, but also that there should be a color spectrum of light.

Quantum Theory Demystified

***Better to adapt the theory to fit the evidence
than to imagine evidence to justify the theory.***

Quantum mechanics is a mathematical theory dealing with the motion and interaction of subatomic particles. It is a very accurate science that has given us atomic clocks, transistors, integrated circuits, lasers, MRI (magnetic resonance imaging), plus our modern communication devices and computers. The predictions of quantum theory work perfectly. It is only the interpretations of experimental facts that raise questions as to what is really going on.

Atoms are solid bits of compressed energy. Exactly how energy is compressed to form the atom, however, appears mysterious – but only because we expect something different from what is really happening.

There are six phenomena at the subatomic level for which quantum theory has questionable explanations: (1) some events appear to be random in that they can be predicted only by probabilities rather than by cause-and-effect, (2) there are quantum jumps, abrupt transitions in particles from one quantum state to another without any apparent cause, (3) there is an uncertainty principle by which either a particle's position or its velocity can be accurately measured, but not both, (4) sometimes pairs of particles are connected (or entangled) in such a way that the quantum state of each particle cannot be described separately, (5) a particle can sometimes be seen in two locations at the same time, depending on how it is observed, (6) the observation of an object appears instantaneously to influence the behavior of another distant object even though there is no physical force connecting the two, and (7) the present behavior of a particle can change its past behavior.

Physics is limited at understanding these known phenomena because the subatomic world may be an interface between nonphysical and physical realities. If so, then science can measure only the physical results of this interaction but not the intermediate transitional states which are not yet fully physical. Whatever is real that science cannot explain must be nonphysical in nature.

Quantum "Probabilities"

Albert Einstein and Max Planck are said to be the fathers of the quantum revolution – yet both were unhappy with the subsequent development of quantum mechanics. Neither were comfortable with the emphasis on probabilities, which elicited Einstein's frequently quoted objection, "*God does not play dice with the universe.*"

Nothing happens by chance. Effect always follows cause. To say that a number of outcomes are possible, each with its own probability, is an admission that one does not know the actual cause of the events in question. Einstein believed that there was something missing from quantum mechanics which, when discovered, would do away with the so-called probability factor. He believed that there must be hidden quantum variables, and that quantum theories would not be complete until those hidden variables were found.

Einstein also objected to the subjectivity of quantum observations because he believed that whatever is real exists independently of how it is observed. He said, "*I think that the particle must have a separate reality independent of its measurements. That is, an electron has a spin, location, and so forth even when it isn't being measured. I like to think the moon is there even if I am not looking at it.*"

Einstein was right about something missing from quantum mechanics, but it is not hidden mathematical variables. Rather, it is the awareness of a nonphysical realm that continuously interpenetrates and influences the quantum world.

Einstein was also right about particles having a reality that is independent of their observation. Subatomic particles do in fact exist, regardless if anyone is looking at them. Observation affects only the various paths that particles may take while in transition from the nonphysical into the physical. All possible paths ultimately have the same destination, however, which is the creation of the fully physical atom.

Quantum Jumps

Atomic electron transition is the change of an electron from one quantum state to another within an atom. This transition appears discontinuous as the electron "jumps" from one energy level (orbital) to another in a few nanoseconds (billionths of a second). In other words, for an extremely small increment of time that electron popped out of existence and then reappeared in a different location. A tiny bit of matter became nonphysical, then became physical again. This is possible only if a nonphysical reality does in fact exist.

Some scientists are not comfortable with the idea that an electron could pop in and out of existence and so use a different terminology. They call an electron "localized" when it can be seen and "nonlocalized" when it cannot.

Quantum jumps create light. An electron has a natural orbital that it occupies. A pulsation of energy causes an electron to jump to a higher orbital. When the electron drops back to its former orbital, the energy it loses is emitted as a photon (packet of light energy). Each element (in the periodic table) has a unique and steady frequency with which electrons jump back and forth between orbits giving off light. In other words, each element has a unique frequency with which its electrons pass back and forth between physical and nonphysical states.

Max Planck developed quantum theory to explain how the vibratory frequency of the electron appeared to be in "chunks" or quanta of energy. Suddenly and without any known cause, an electron would suddenly radiate a quantum of energy as a pulse of light. X-rays and other electromagnetic waves can be given off only in these discrete packets/quanta. Neils Bohr also talked about light being emitted in quantum jumps. [*A quantum jump is the change in energy by a single quantum.*]

There is a Copenhagen interpretation which postulates that subatomic events are only perceptible as indeterministic physically discontinuous transitions between discrete stationary states. This is a long-winded way of saying that there are tiny increments of time during which these events are nonphysical ("physically discontinuous"), flickering in and out of existence.

Stephen Hawking has remarked that *"quantum fluctuations lead to the creation of tiny universes out of nothing"*. If we translate "nothing" in this context to mean a nonphysical reality, then Hawking's observation is most fitting. Nothing cannot create anything, but energy can. If the nonphysical is a state of pure primary energy, then it can create not only tiny universes but also the entire universe.

Uncertainty

Subatomic particles don't appear to have separate, well-defined positions and velocities. Instead, they have a quantum state, which appears to be a combination of position and velocity.

Werner Heisenberg formulated the uncertainty principle whereby the more accurately you measure an object's position, the more uncertain you will be about its momentum – and the more accurately you measure its momentum, the more uncertain you will be about its position. This uncertainty appears to be caused by quantum fluctuations, temporary changes in the amount of energy in a point or particle in space.

For Heisenberg, this uncertainty raised a philosophical issue and suggested that *"atoms or elementary particles themselves are not real; they form a world of potentialities or possibilities rather than one of things or facts."* In other words, atomic-scale objects exist only in nonphysical reality and not in the physical world.

It is helpful to view subatomic particles as fluctuating energy packets in transition from the nonphysical into the physical. At times these pre-physical particles may have a position, and at others a velocity. When you intend to measure where it is located, the particle co-operates by letting you know exactly where it would be if it were to become fully physical in that moment. If you intend to measure how fast it is moving, the pre-physical particle lets you know the exact velocity at which it can travel in physicality. In other words, the position and velocity of the pre-physical particle are potentialities rather than actualities – neither of which has any bearing on the final result (the atom), which is always the same regardless of how its component particles may behave.

Connectedness

There are some repeatable phenomena observable at the quantum level which strongly suggest universal connectedness. Simply stated: any two objects that have ever interacted may be forever connected. The behavior of one instantaneously affects the other and the behavior of everything connected with it, no matter how far apart they may be.

John Bell viewed connectedness in terms of non-separability. His perspective was that it is impossible to separate objects so that what happens to one in no way affects what happens to others. Without separability, what happens at one place can instantaneously affect what happens far away, even though there is no physical force connecting the objects in question.

Some refer to universal connectedness as "entanglement" – a physical phenomenon that occurs when pairs of particles are generated or interact in ways such that the quantum state of each particle cannot be described independently. When two of these particles become separated, the state of one affects the state of the other. One particle of an entangled pair appears to duplicate or mirror instantly whatever measurement has been performed on the other, even though the two particles may be separated by large distances. If you manipulate what is going on in one location, you automatically change what is going on in the other location, even though there is no physical force connecting them and no time for information to pass from one to the other. There is obviously a direct energy connection of some kind between two entangled particles for this to happen.

The experimental facts concerning entanglement are convincing evidence of a nonphysical reality. The logic is thus: (a) two objects are in fact connected, (b) there is no physical force between these two objects, thus (c) whatever is connecting them has to be nonphysical.

The time between the detection of one photon and the detection of its twin is less than the time it can take for light to go from one detection apparatus to the other. Since light is the ultimate limiting velocity in physical reality, whatever communication is happening between the twin particles must be nonphysical – i.e., at the speed of thought.

Experiments by Alain Aspect during the 1980s suggest that communication between particles 12 meters apart takes place in less than one billionth of a second. Experiments by Nixcolas Gisin in the 1990s suggest that particles 10 kilometers apart appear to be in communication 20,000 times faster than the speed of light.E

Suppose twin particles are not directly connected to each other but are two different individuations simultaneously expressing from a common source.

> **Thought Experiment**: Imagine that you have a laser pointer pen on which is fastened a beam splitter so that it projects two identical dots onto a screen. Whichever way you move the laser pen, the two dots would move in unison in precisely the same direction at the same exact time. It would appear as if the two dots were instantaneously communicating with each other, but it is really a single pen projecting the identical image in two places simultaneously – making the two dots move in response to your intention. Perhaps this is how entanglement works: twin particles are dual projections of the same nonphysical intention.

The speed of thought is instantaneous, infinitely faster than the speed of light. Perhaps the two particles are manifestations of the same thought. Although lacking in consciousness themselves, the particles respond to mental energy that originates from a conscious source.

In Two Places at Once

Experiments by Niels Bohr, Erwin Schroedinger, and others show that particles, such as electrons, can be in two places at once and can also behave as a particle in one moment and as a wave in the next, depending on how an observer decides to measure it. These apparent anomalies occur only because the particle in question has no definable state or position until such are measured.

Experimental evidence shows that subatomic particles can at the same time be in two different positions, travel on two different paths, have two different polarities, and/or function as both a particle and a wave. The particle in question is said to be in a superposition state, meaning that it exists in all possible states simultaneously provided we don't observe or measure it. It is the act of measuring that causes the subatomic particle to express itself through only one of these possibilities.

Existing in all possible states (superposition) happens when the subatomic particle is in a pre-physical state capable of taking all possible paths on route to becoming fully physical. When we decide to measure a specific aspect of this pre-physical particle, it takes a specific route into physicality. The superposition state is then said to collapse because all the other possibilities have been replaced by a single actuality.

It is the experiment that determines which of two possibilities is observed. Your choice of measurement will determine whether you see (a) either a particle or a wave, (b) either a position or a velocity, or (c) a particle in either one location or another. As an observer, you do not see whatever you wish. Your choice is limited to seeing which of two properties (or locations) a quasi-particle will manifest as it becomes physical.

According to the Copenhagen interpretation, one could project a single proton in its superposition state into two different boxes. When the boxes remain sealed, the proton would exist as a potentiality in both locations. Then, when the door of one box is opened, the proton appears in that box and the other box is empty, regardless of which box is the first to be opened. There is an inconsistency in this theory. It is impossible to project a superposed proton into a sealed box because the moment it hits the exterior surface of the box it has become physical and stops there – just as one can project protons onto a screen but not through that screen.

Some theorists took the idea of an electron or proton appearing in two places simultaneously to the next level, to assume that a complete atom (or an object) can exist as a combination of multiple states corresponding to different outcomes. This is an impossibility. An atom is fully physical and cannot exist in a superposition state (which term applies only to pre-physical possibilities).

Erwin Schroedinger dismissed the false assumptions of the Copenhagen interpretation by showing that its logical consequences are nonsensical. By means of his infamous cat-in-the-box thought experiment, Schroedinger demonstrated (tongue-in-cheek) that a cat could simultaneously be both dead and alive. The imaginary experiment consisted of putting inside each of two sealed boxes (a) a flask of poison, (b) a tiny bit of radioactive substance so small that one atom of it has a 50:50 chance of decaying within one hour, (c) a Geiger counter that when it detects decay, triggers a hammer to break the flask of poison. A cat in its superposition state is projected into both boxes. If no decaying atom is detected in either box, then a live cat will show up in whichever box is opened first and the other box will be empty. If, however, a decaying atom is detected in one box, the cat in that box will be killed. As long as the boxes remain sealed, the same cat will be alive in one box and dead in the other.

There are three fallacies that the cat-in-the-box *reductio ad absurdum* experiment reveals: (1) because cats are physical objects, they cannot have superposition states and thus cannot be projected into multiple locations, (2) mutually exclusive conditions cannot exist in superposition states (e.g., a particle cannot be a proton at one location and not a proton at another), and (3) observation does not alter outcomes; the experimental method we choose determines only which aspect of reality we see but not what that reality ultimately is.

Observation

Experimental evidence suggests that an observer may influence the behavior of subatomic particles. This is not entirely accurate. It is the choice of experimental measurement that determines whether one sees a particle or a wave, a particle's position rather than its velocity, a particle in two places rather than one, or a particle that has had one history rather than another. The observer chooses his experience of reality, but not reality itself. An atom is an atom, and the moon is the moon, whether they are being observed or not.

If a physical property of an object can be known without its being observed, then that property could not have been created by observation. It must have existed as a physical reality before its observation. The early universe existed long before there were any witnesses to observe it.

The choice of experiment affects what is observed only because subatomic space is a transitional state in which nonphysical energy is in the process of becoming physical. It can do so in multiple ways, and it does not matter to the result (the fully physical atom) which interim path is chosen. If an observer wishes to see a specific aspect of the transition, nonphysical reality obliges.

Human observers have thoughts – and thoughts, being nonphysical, are communicated instantaneously. By structuring an experiment in a specific way, the observer instantly makes her intention known to the nonphysical source of the single or twin particles, which source co-operates with the experiment. Whereas the energetic phenomenon in question is always there, it is not always in the same presentation.

A photon duplicates its twin's behavior instantaneously. Because there is no time lapse, it isn't possible for information to be communicated by any physical means from the observer to the particles. Whatever is going on for the particle and/or its twin exists independently of anyone's observation of it or them. This meets Einstein's analogous criterion that the moon is still there when we are not looking at it.

Multiple "Histories"

The quantum state is not an objective physical thing. It is rather an intermediate stage of transition between nonphysical and physical realities. The nonphysical aspects of this transitory state include many possible paths that particles can take on route to their destination. Only one of these possible paths ultimately becomes the real (physical) path, depending on the intention of an observer. If no observer is present, then the particles may take different paths – and none of this makes any difference to the result. All possible paths lead to the same destination, the creation of the identical physical atom.

When physicists fire particles (single photons or electrons) through two very narrow slits onto a photographic plate when no one is observing, these particles form a wave interference pattern. When someone is observing, however, the particle almost always behaves as if it is a singular entity, seemingly choosing one of the two slits through which to pass rather than through both. The act of observing appears to select the path the particle takes. Some kind of mind-matter interaction is taking place.

Richard Feynman (*1918-1988*) came up with a plausible explanation for this phenomenon. He demonstrated that a particle going from A to B simultaneously appears to take every possible path in space and time to get there. It is as if the particle has no definite position between the start and end points until it is called into action for a specific purpose. It is as if the particle does not have a single history but many possible histories.

The concept of multiple histories is akin to the expression, "All roads lead to Rome." None of these metaphorical roads change the atom itself, however.

It is the experiment that appears to select which of the possible paths (or histories) that the particle takes on its way to its destination. In our physical world, there is a time lapse between beginning on a path and ending at a destination. However, in the nonphysical quantum world in which time does not exist, there is no such thing as a beginning, no such thing as an ending, and no such thing as a history. In subatomic space, time is an illusion. Feynman's brilliant insight about all possible paths existing simultaneously is really an observation about infinite possibilities for manifesting (i.e., bringing into physical reality).

Hypothetical Particles

Scientists detect temporary, fleeting pulsations of energy at the subatomic level and assume that such blinks of energy must be caused by physical matter of some kind, e.g., undetectable particles. They postulate that there must be hidden virtual particles that occur over a very short interval whose supposed existence accounts for the measurable effects being observed. The term, "virtual", is a tipoff. It literally means "recognized as such but not strictly so". There is no proof for the existence of virtual particles, which are literally defined as having only the appearance of reality.

A virtual particle is an artificial device used to fill a mathematical gap or make an equation balance. If the result of a subatomic process is known and only one of the two interactive forces required to produce that result is known, then it is assumed that the unknown force must be caused by a hidden particle of some kind.

Virtual particles are (i) undetectable, (ii) travel faster than light, (iii) can progress back in time, and/or (iv) may violate the conservation of energy principle. No physical bit of matter (e.g., actual particle) has these characteristics, all of which defy the principles of physics. Therefore, whatever unknown force is at work must be nonphysical in nature. Nonphysical energy does not require a physical cause (e.g., a particle). Nonphysical energy simply is.

String theory is (or was) an assumption that virtual particles can be described as patterns of vibration that have length but no height or width, like waves travelling down infinitely thin pieces of string. The premise upon which string theory rests is faulty: anything which has zero height or width must also have zero length. It is impossible for these hypothetical "strings" to have any physical existence whatsoever. Whatever this theory is attempting to describe can only be nonphysical energy in transition to a physical state.

Antiparticles are virtual particles hypothesized to exist to explain a force of attraction that isn't otherwise understood. An antiparticle supposedly has the same mass as a given particle but opposite electromagnetic properties. The assumption is that if a known particle is behaving as if another particle is attracting it, then that second particle must exist even if it cannot be detected. [*Another case of imagining evidence to fit the theory, rather than modifying the theory to explain the evidence.*]

If one makes the limiting assumption that all the energy within an atom is fixed, then it is only natural to look for physical causes for everything that happens within that atom. By this questionable reasoning, all energetic changes within the system must be caused by some form of matter that is already present. The possibility that energy could be entering this physical system from an outside source (i.e., nonphysical) is not even considered. [*The equivalent error in cosmology is assuming that that the total energy in the universe is fixed; therefore, unexplained energetic changes must be caused by hidden "dark" matter.*]

Particle-as-cause assumptions lead to questionable conclusions – such as empty space within the atom is filled with virtual particles and pairs of antiparticles that together have an infinite amount of energy. "Infinite" is a tipoff in this hypothesis. Only a nonphysical source can provide infinite energy.

There is a widespread assumption that the hidden variables that Einstein sought can only be explained by the positions of hypothetical particles. There is another more plausible explanation: both hidden variables and virtual particles are flawed attempts to explain transitional nonphysical energy in purely physical terms.

If we consider subatomic space to be a dynamic interface between nonphysical and physical realities – then quantum jumps, the uncertainty principle, and all other peculiarities of motion at the subatomic level can be explained by energy that is in transition from the nonphysical to a differentiated physical state. However temporary or fleeting this transition may appear, nonphysical reality always provides the energy required to maintain the homeostasis of the atom, in whatever form it may be required.

Nonphysical energy manifests into subatomic space in a pure and undetectable state, as a potentiality. This primary energy takes on physical properties only when it differentiates itself for purpose. It is only during this differentiation process that our instrumentation can detect its presence, which often appears fleeting and transitory. [*There is a parallel to this process in biology, where an undifferentiated stem cell with certain general characteristics has the potential to become any number of specialized cells, each with its own unique characteristics.*]

Nonphysical energy continuously interfaces with subatomic space and maintains the homeostasis of the atom. If this nonphysical energy were suddenly to be withdrawn, all atoms, all matter, and the whole universe would collapse. Therefore, what we are witnessing at the subatomic level is the continuous recreation of every atom in the universe. What we observe as physical matter is ultimately a continuous projection of nonphysical reality into the physical realm.

Life Emerges

Chemistry cannot create biology.

Life is assumed to have emerged spontaneously by means of abiogenesis, the formation of living organisms from non-living substances. This term is simply a label which identifies the phenomenon but tells us nothing about its cause. How could abiogenesis be possible?

In the early 1950s, Stanley Miller and Harold Urey attempted to create life from nonliving chemicals. They simulated primordial conditions of some four billion years ago, when life on earth appeared for the first time. Their experiment involved water, methane, ammonia, hydrogen, evaporation, condensation, and firing continuous electrical sparks to simulate lightening. What Miller and Urey were able to produce was a smelly sludge containing some inert amino acids – but no life.

Amino acids are the building blocks of proteins which become essential components of the tissues, hormones, enzymes and antibodies of animals and humans. All amino acids are formed from four elements: hydrogen, oxygen, nitrogen, and carbon. Some amino acids also include sulfur. All five of these elements are widely distributed throughout the universe.

Miller and Urey thus demonstrated that natural conditions can create inert biological raw materials, not only on earth but perhaps also on billions of other planets and maybe even on rocks hurtling through space. Some of the meteorites that have crashed to Earth have been found to bring with them sludgy amino acids.

Scott Sandford, a NASA scientist, built an artificial comet in a laboratory by recreating the atmospheric conditions of deep space. To methanol, water, and carbon dioxide he added ultra violet radiation – and the result was the creation of primitive amino acids.

Inert amino acids are not life, however. The Miller-Urey and Sandford experiments thus demonstrated that non-living things cannot create life – that chemistry alone is incapable of creating biology. Metaphorically speaking, something has to breathe life into inert chemicals in order to create living things. The logical candidate for this "something" is nonphysical energy in a differentiated form called "life energy".

It is doubtful that scientists could ever create life from inert amino acids. This is because every life form requires a DNA program to assemble the proteins that make up its structure, and to create and maintain the metabolism required for its functioning. How would scientists introduce DNA into the experimental mix? The only available source of DNA is from other living organisms, in which case the experiment would become one of altering life rather than creating life.

The inert amino acids created by primordial conditions thus lack two prerequisites to becoming living organisms: (1) life energy, and (2) DNA, both of which would have to be pre-existing. At some point in time, on this and every other planet capable of sustaining life, the nonphysical must have made deliberate decisions about how and when to create the life forms most suitable to every given situation.

Chemistry cannot create biology. Only the nonphysical can, when conditions and timing are right.

Remarkable DNA

All living creatures require deoxyribonucleic acid (DNA), the self-replicating genetic material present in their cells as a component of chromosomes. DNA carries elaborate and precisely encoded instructions for cell reproduction.

Each DNA molecule consists of a base pair of nucleotides, either guanine (G) coupled with cytosine (C), or adenine (A) coupled with thymine (T). Human DNA consists of a chain of about three billion of these GC and AT base molecules linked together. If only two or three out of 1.5 billion DNA molecules are out of sequence, birth defects or congenital disease can be the result. If only five to 10 of these 1.5 billion molecules are defective, death can be the result. This is an error factor of less than 0.00000001 percent. Somehow, our DNA has been programmed with uncanny precision.

GC and AT base molecules linked together in long chains is analogous to binary computer coding. Mechanical computers are programmed in a binary machine language in which each digit is either a "0" or a "1". Living cells are similarly programmed in a binary language in which each molecule is either a "GC" or an "AT". Ten binary codes in sequence make possible $2^{10} = 1,024$ unique combinations. Three billion binary codes in sequence (as in humans) make possible $2^{3,000,000,000}$ or $(1,024)^{300,000,000}$ unique combinations, which is a number so incredibly huge that it might as well be infinite for all intents and purposes.

Every life form has its own DNA program. Surprisingly, the simpler the organism, the longer its DNA chain can be. The single celled amoeba, for example, has about 300 billion nucleotide pairs in its DNA chain, compared to only 3 billion pairs in human DNA. The amoeba, however, uses only an infinitesimal fraction of all the DNA that it carries within it.

Why does the amoeba have 100 times as much DNA as humans? The answer is twofold: (1) every organism carries within its body DNA potential for possible use by subsequently evolved species, and (2) the more evolved an organism becomes, the more of its ancestral DNA is shed as no longer being required.

Every species actively uses only a tiny fraction of its DNA. For example, only about 1.5 percent of human DNA may be active. The other 98.5 percent or so is inactive. Nature must have a reason for continually and consistently reproducing so much inactive DNA in all species. The reason is that it provides potential for species to adapt and evolve. Each species comes with a genetic imprint of what it could evolve to as a species, plus endless possibilities for the evolution of new species.

Humans and chimpanzees have DNA that may be about 99 percent the same. What makes us different from them is that more of our DNA is active. Chimps may be using only about 70 percent of the same DNA that we are using. The ability of a species to tap into its DNA reserves gives it the potential to take major leaps forward in evolutionary development.

We humans also have about 50% of our DNA in common with bananas. This makes sense only if bananas are carrying most of their DNA as unused potential as a contingency plan in case of mass extinctions. Little or none of the DNA that bananas have in common with us is in actual use by them. If it were, one would expect these two species to have more physical traits in common. Fruits and primates are about as different as two species could possibly be.

Atrophy of Disuse

DNA that is no longer required atrophies from disuse and is lost to future generations. We know this from our human experience with vitamin C.

All mammals, with four exceptions, manufacture ascorbate (vitamin C) in their livers. The enzyme, L-gulonolactone oxidase, converts glucose (blood sugar) into ascorbate, a liver metabolite that serves many functions – including protecting blood vessels, skin, bones, teeth, and gums – as well as warding off infections and ameliorating stress. Animals that produce ascorbate internally are resistant to heart attacks, strokes, osteoporosis, dental decay, gum disorders, and the common cold.

The four mammals who have lost the ability to produce ascorbate are humans, apes, guinea pigs and fruit-eating bats. The reason they have lost this ability is that their ancient ancestors lived on a plant-based diet that was so rich in vitamin C that their bodies no longer had to produce any internally. The part of their DNA that enabled their bodies to produce L-gulonolactone oxidase was no longer required, dried up, and could no longer be passed on to future generations.

The atrophy of DNA responsible for ascorbate production explains why advanced life forms have so much less DNA than primitive single celled organisms like the amoeba. Multicellular fungi no longer need the DNA required to replicate unicellular organisms, and so they lose it. The amphibian that evolved from a fish no longer needs those parts of its DNA that were exclusive to fish and so loses them. Similarly, the lizard loses those parts of its DNA that were required by amphibians ... and so on up the evolutionary scale.

Evolutionary Reserve

Shedding DNA that is no longer required implies that all DNA retained by an organism must serve a useful purpose, otherwise it would not be preserved. Only a tiny fraction of DNA is currently active and being used by the plant or animal itself. The only conceivable purpose served by carrying forward such disproportionately huge amounts of inactive DNA is for creative potential, to keep biological codes in reserve for all possible creatures which could evolve from the present ones. This orchestration not only provides for limitless development and expansion of existing life forms, it also includes backup contingencies in case of mass extinctions.

About 250 million years ago, there were major volcanic eruptions that continued for some two million years. Ninety percent of marine animals and 70 percent of land animals were wiped out. The survivors evolved into dinosaurs, which dominated the planet for about 165 million years. Without this cataclysmic event, there would have been no dinosaurs.

About 66 million years ago, a 10 trillion-ton asteroid over six miles in diameter (about the size of Mt. Everest) slammed into Earth, leaving an impact crater over 100 miles wide that is now buried under the Yucatan Peninsula in Mexico. Over 70 percent of all species, including the dinosaurs, vanished because of this asteroid collision. Those not killed by the impact and its fallout perished because dense clouds of debris blocked the sun, thus halting photosynthesis and starving the life forms dependent upon it. Alligators, crocodiles, frogs, salamanders, and spiders survived; but large land animals did not. Evolution was dealt a serious setback but rebooted, thanks to the inactive DNA potential of the surviving species.

In the next 10 million years following this asteroid disaster, every major animal group that is around today burst onto the scene. There was a prolific divergence of life into new forms and species that had never existed before

– including flowering plants, birds, and large mammals (eventually including ourselves) – all made possible by calling into play unused DNA potential that had been waiting in reserve for just such a contingency. The death of species became the birth of new species.

The only mammals that appear to have existed prior to the asteroid collision were rodents that served as a food source for dinosaurs. It was only after the dinosaurs became extinct that the surviving mammals evolved into more advanced forms, including primates and humans. We owe our existence to the extinction of the dinosaurs.

If an even greater disaster had wiped out all life forms except for unicellular organisms, evolution would have recreated itself. Amoebae, algae, and fungi may contain enough inactive DNA coding to create every species that have ever lived or ever could live on Earth.

DNA and Evolution

Evolution is impossible without DNA. Therefore, whatever created DNA also created evolution, and vice versa. Either DNA was the result of random events or it was orchestrated. There is no third possibility. All the evidence indicates that DNA coding could not have been random.

DNA structure and function are too complex to be explained by known evolutionary mechanisms. The DNA biological system could not have evolved by successive small modifications to pre-existing systems through Darwinian natural selection or random mutation.

Random mutation rates are very low; and most mutations are harmful, making an organism less capable of surviving. Furthermore, DNA copying processes have sophisticated built-in repair mechanisms. In the rare instances where genetic mutations may be beneficial, they can make only minor changes to organisms and are incapable of developing new body plans. The birth of new species thus requires pre-planning.

There is no possible way that programmed DNA could be the result of random events. In the case of the amoeba, it took foreknowledge – to direct (a) cytosine to combine consistently with thymine, (b) adenine to combine consistently with guanine, (c) 300-billion of these CT and AG base pairs to line up in a precise binary sequence, (d) the arrangements of these binary codes to be consistent with precise logic and nontrivial computing – to create (e) a double helix structure in which an AG purine in one strand bonds to a CT pyrimidine in another strand, and vice versa – and to build in (f) self-correcting polymerase enzymes that proofread their work at each stage of DNA development, and (f) surgical and chemical ways of correcting DNA damage after the fact.

There are three questions that randomness theory cannot answer:

1. Once billions of DNA molecules had randomly lined themselves up in precise sequence, how did they know what they were supposed to do?

2. Why would random events produce billions more DNA molecules than required for the life and maintenance of any given organism?

3. How could random events know which DNA molecules to put in reserve for the evolution of new species?

There is another difficulty with the "DNA just happened" argument. DNA is an essential component of life. Before there was DNA, life did not exist; there were only physical and chemical substances. If the four nucleotides involved in DNA were spontaneously created by means of chemical reactions in multiple locations at various times, how did they seek each other out to form DNA? There is no electromagnetism, no chemical affinity, nor any other physical force drawing them together. Only living organisms are capable of the independent motion required to seek out each other, but nothing was living until there was DNA. There appears to be a hidden false assumption in the spontaneous DNA argument, namely that of ascribing lifelike characteristics to nonliving chemicals.

Life is impossible without DNA. DNA is required to create the proteins from which cells, tissues, organs, hormones, enzymes, and antibodies are formed. For DNA to function, however, it requires the pre-existence of enzymes.

The DNA polymerase enzyme is a protein molecule comprised of over 700 amino acids. Amino acids are incapable of spontaneously combining to form proteins. The enzymes required to enable them to combine can only be created by means of some biological (i.e., living) method of protein synthesis. This means that the very first living organism capable of synthesizing proteins would not have been able to function without the prior existence of proteins synthesized by some other form of life. There are only two ways out of this dilemma: either (1) some other form of life pre-existed DNA based life, or (2) both life and DNA were created by nonphysical reality.

Advocates of randomness sometimes refer to the infinite monkey "theorem" as their authority. This speculative theory claims that a metaphorical monkey hitting keys at random on a typewriter keyboard for an infinite amount of time will almost surely type out all of Shakespeare's plays. "Almost surely" is a tipoff that there is no guarantee that this could ever happen. Even an infinite number of monkeys typing even one Shakespearean play has a probability so small as to be negligible (but technically not zero). Can you imagine checking in on this experiment 400 trillion years later to find that one monkey managed to type the first scene of *Hamlet* before lapsing into complete gibberish? If the creation of DNA were a random event, perhaps the universe would still be waiting for it to happen.

To understand DNA, we need to return to our computer analogy. Every mechanical computer requires two levels of programming: (1) the creation of a mechanical language (binary code), and (2) instructional programs detailing the calculations and decisions the computer is to make in order to accomplish specific objectives. So is it also with DNA. Both mechanical computers and biological computers require these two levels of programming.

Many scientific minds realize that DNA is so complex that it could not possibly have evolved on planet Earth. This conclusion leads to speculation that DNA life forms may have come to us by means of (1) panspermia (via meteoroids, asteroids, comets, planetoids), and/or (2) directed panspermia (via alien visitors). Both theories could be true; however, they do not explain how DNA could have evolved on other planets. DNA is so complex that it could not have evolved anywhere. It had to have been created.

In support of the panspermia theory, Professor Mark Burchell simulated the conditions of a comet hurtling into sterile water at 14,000 miles per hour. Aboard this projectile were live plankton, a small percentage of which not only survived the impact but also proliferated, effectively creating a new colony.

The only explanation which fits the evidence is that a universal mind must have created the DNA template, programmed it, introduced it into the most primitive of life forms everywhere, and let evolution take it from there. Once the DNA foundation is in place, species evolve by selective adaptation to their environments and by mating with partners that have different characteristics, both of which methods produce offspring with DNA that is different from their parents. Natural selection requires pre-existing DNA for it to work.

Darwinian evolution presupposes that the development of more advanced life forms can happen only from a combination of these two factors: (1) random mutation, and (2) natural selection. There is another factor overlooked by evolutionists: (3) cellular adaptation. Cells can acquire new characteristics resulting from interaction with their environment, and these characteristics are inherited by successive generations. In other words, cells can learn and adapt within their own lifetime, making significant changes without having to wait for them to show up in their descendants.

Cellular adaptation is most probably a more significant factor in evolutionary processes than natural selection. Animals and plants have built-in abilities to change to the expression of their genes – by switching on inactive genes and switching off active ones. There have been periods in our planet's history when the proliferation of new species far outstripped the ability for them to have been created only by natural selection. These are most probably epochs during which cellular adaption significantly accelerated evolution – the most noteworthy one being the 25-million years of the Cambrian explosion, during which all the body plans which currently exist in all animals suddenly (in evolutionary terms) appeared.

Cellular adaptation is facilitated by transposons, which are mobile segments of DNA that can replicate and insert copies of other DNA segments at pre-selected sites in the same or a different chromosome. Transposons thus alter the genetic constitution and expression of an organism.

There are two questions about evolution that beg for answers: (1) How was DNA created, and (2) Why does every organism always pass on to its descendants billions of inactive DNA codes whose only plausible function can be to serve as untapped potential for new species? Natural selection takes us only as far as understanding how two parents can have offspring with characteristics of both parents, and that the resulting combined DNA may make the children better adapted to their environment than were succeeding generations. However, natural selection is not intuitive. It cannot predict which DNA codes may be required by future generations and future species. Only a universal mind can know such things.

A Universal Template

A universal DNA program in which each species accesses only its own sub-module provides unlimited potential for the adaptation of every form of life. This kind of programming ensures that every species carries in its huge reserve of inactive DNA some unused sequences that are common to many other species, both present and future. It makes possible adaptive evolution through the survival of specific genes, rather than the survival of select species. It explains similarities in species that have completely different origins. Two unrelated species can, on separate evolutionary pathways, activate some DNA potential common to both.

No species could either exist or reproduce without its specific sub-program that fits within a universal DNA template common to all life forms. This may be so not only on Earth but perhaps also throughout the universe. Estimates suggest that there may be 40 billion Earth-like planets in the Milky Way galaxy. Wherever life may exist in any of these other worlds, it would most likely have followed the same pattern as here, from the same raw materials, the same evolutionary patterns, and with the same universal DNA programming uniquely adapted to each situation.

Because environments differ greatly within habitable zones, we would expect many species of life on other planets to be quite unlike anything on Earth. Most of the life forms which thrive on other planets may not survive if transported here, and vice versa. However diverse and varied interplanetary species may be, they all require DNA programming of some kind. It is not inconceivable, however, that some very specific alien life forms could have DNA modules that may be compatible with their earthly counterparts. Perhaps there are marine animals or even humanoids elsewhere that could interbreed with closely related species here.

The universal DNA template holds possibilities for improving the genetics of humans, as a species. Scientists do not need to create artificial DNA to accomplish this. All they need do is to examine the 90 percent of our DNA that is inactive, to see if the characteristics they seek are already there.

Evolutionary Anomalies

***There are gaps in evolutionary theory
that science cannot explain.***

How did life evolve from non-living organic chemicals? Did it happen by chance or by intention? Here is what Fred Hoyle had to say on the subject: *"Life cannot have had a random beginning ... The trouble is that there are about 2,000 enzymes, and the chance of obtaining them all in a random trial is only one part in $10^{40,000}$, or an outrageously small probability that could not be faced even if the whole universe consisted of organic soup."*

The fact that many species share the same physiology, biochemistry and cellular composition does not necessarily mean that all have descended from a common ancestor. The organ cells of all mammals are the same in structure and function, regardless of the animal from which they come. In other words, the cell from a pig's heart (or a sheep's, or a cow's) is essentially the same as the cell from a human heart. Does this mean that pigs and humans have a common ancestor? Not necessarily. All it really means is that for a heart to work, it must be composed of cells that are anatomically correct and function with the appropriate physiology, regardless from which ancestral lineage the animal comes. The same holds for mammalian cells of the lungs, kidney, prostate, thymus, spleen, thyroid, liver, adrenal glands, etc.

If mammals have evolved on other planets in atmospheric and gravitational conditions similar to ours, the cells of their internal organs would be essentially the same as those mammals which have evolved here, simply because that is how evolution works. Similarly, if fish or reptiles have evolved on other planets in sea water that is of similar composition to ours, the cellular physiology of such creatures could be the same as those living here, even though there may be significant differences in anatomical structure and outward appearances.

Cellular Evolution

About 3.5 billion years ago, the first life forms to emerge on earth were prokaryotes --- single celled microorganisms with DNA but without membranes or nuclei. The most abundant life forms on earth today are prokaryotes, of which bacteria and blue green algae are examples.

These single-celled organisms have complex protein manufacturing processes –
and a sophisticated cell control system that could not have evolved in small steps, because it can only work as a whole system. To reproduce themselves, prokaryotes must have had all these necessary systems in place from their very beginning of life on Earth.

About two billion years ago, eukaryote cells came on the scene. The eukaryotic cell is different from bacteria and algae because it has a more intricate structure that includes many new constituents – such as a nucleus, mitochondria, chloroplasts, and other organelles. Eukaryotic cells are larger, more complex, and have larger DNA and more intricate processes than prokaryote cells.

The main feature of the eukaryotic cell is its ability to form advanced multicellular organisms, including plants and animals. It became the building block of all higher forms of life, including humans. The human brain, skin, muscles, skeleton, and all internal organs are built from this same type of cell adapted to perform specific tasks. Because all cells in the body are essentially the same, they can cooperate with each other as one entity.

It is generally believed that at a certain point in time, one prokaryote cell entered another and became its nucleus to create the very first eukaryote cell, from which all higher life forms have descended. This merging of two prokaryotes was a supposedly random event that happened only once; and the resulting prokaryote is attributed with being the single ancestor of all fungi, plants, animals, and humans now living on our planet.

Random or not, this merging of two prokaryotes must have been anticipated in their DNA programming. One cell entering another with the identical structure does not explain how unique new cellular components (e.g., nucleus, cytoskeleton, Golgi apparatus, etc.) were created that could not have been taken from the progenitor bacteria that did not have such structures. It could happen if and only if these future contingencies had been preprogrammed into the unused, evolutionary reserve DNA of the prokaryotes in question.

Eukaryotes developed in two main directions. Some became animal cells. Others acquired cell walls and became plant cells containing chlorophyll and starch. All animals and plants may thus be considered descendants of eukaryotes that combined to form complex organisms.

Two prokaryotes merging to form a eukaryote may have been unusual; but this was not a singular event, nor was it random. Nothing happens by chance. Forces of attraction are constantly at work. Prokaryotes come with the DNA potential to form eukaryotes whenever their doing so fits within a larger plan. Newly created eukaryotes may happen on rare occasions intermittently over periods of millions of years.

As prokaryotes, bacterial cells have approximately 10 percent of the DNA base pairs that humans have. When two prokaryotes merge, one would thus expect the resulting eukaryote to have about 20 percent of the DNA that we do. However, the eukaryotic amoeba has 100 times the DNA of humans – or 500 times as much as it could have inherited from its ancestral prokaryotes. This leads us to the inescapable conclusion that a huge amount of DNA must somehow have been added to primitive life forms in a way that has nothing to do with evolution. It also draws into question the origin of DNA itself. The very first prokaryotic cell had no ancestor from whom it could have inherited DNA.

Origins of Species

There is a centuries old clash between creationism and evolution. Creationism is the theory that attributes all species to separate acts of creation. Evolutionary theory maintains that simple life forms, over extended periods of time, gradually develop into more advanced species. Charles Darwin, proponent of evolution by natural selection, had hoped that one day it would be possible to trace every life form on the planet back to a single ancestor. If such were the case, then how did that original ancestor get here? There were no prior species from which it could have evolved; therefore, it must have been created. If one was created, then why not many?

In South America, Darwin discovered fossils of extinct creatures whose structures bore strong similarities to animals presently living in the same area. He theorized that the modern animals must be descendants of the extinct ones, and that somehow species are able to evolve into new species.

When Darwin reached the Galapagos Islands, he found variations of birds and reptiles that differed from those on the mainland and also differed from island to island. Mockingbirds had different shaped beaks uniquely adapted to the food sources on each of the islands. Finches had also adapted, so much so that those on one island were incapable of mating with finches on other islands. Darwin reasoned that through a process he called "natural selection", successive generations of animals modify their bodies to adapt to new living conditions. This they do by a kind of survival of the fittest: those offspring who inherit characteristics better suited to their environment survive, thrive and multiply – whereas those who inherit less suitable characteristics die off.

Worldwide there are about 40 species of birds that, by natural selection, have evolved the inability to fly. A typical example is a fish-eating bird that flies from a mainland where fish are scarce to a remote area where fish are plentiful. In this new environment of abundant food, this bird no longer needs to fly to catch fish. It simply dives into the water to feed itself. The less this bird flies, the more its wing muscles atrophy from disuse and the less capable it becomes of flying over any appreciable distance. Each generation of offspring are born with wings less and less capable of flight, until one of these generations can officially be classed as a new species of flightless penguin.

Natural selection is a very slow process. Darwin speculated that it took "eons" to produce new species. As an example, the wolf (*canis lupus*) that roamed the earth about a million years ago became ancestor to the jackal (*canis aureus*), the coyote (*canis latrans*), the dingo (*canis lupus dingo*), and the dog (*canis familiaris*). Because jackals, coyotes, dingoes, and dogs can all interbreed, perhaps they should all be considered specialized wolves rather than new species.

Artificial selection is a very rapid process, a fast forwarding of cellular adaptation by which humans have created some 350 breeds of dogs plus countless mongrels of widely varying sizes, shapes, colors, sensory acuity, capabilities, and temperaments. All dogs, from the tiny Chihuahua to the huge Great Dane, however, are members of the same sub-species of wolf – and thus demonstrate the limitless genetic potential from which every species is ultimately able to draw for its own development. Every wolf must have in its untapped DNA the possibility for its descendants to become poodles or boxers or every kind of dog imaginable. These 350 breeds of dogs have nothing to do with evolution, however, because they do not generate new genes. They simply rearrange possibilities within the spectrum of the existing wolf genes.

The essence of evolutionary theory is that over hundreds of millions of years, life began in the sea, fish evolved into reptiles, some of which evolved into dinosaurs, some of which evolved into birds. Some cold-blooded reptiles also evolved into warm blooded mammals, including the primate family. Some of these land mammals returned to the sea in the form of whales and dolphins.

How did fish end up walking on land? The mudpuppy of Ontario, Canada may be one example. The mudpuppy is a completely aquatic salamander that has external gills and four legs, each with four toes. Salamanders may be the bridging species between fish and land reptiles.

Charles Darwin believed that all species of life have descended over time from common ancestors. He was unable to realize his dream that it would be possible to trace every life form on the planet back to a single ancestor, however. The closest he could come was, *"I believe that animals have descended from at most only four or five progenitors and plants from an equal or lesser number."* This is an admission that there must have been multiple acts of creation of some kind.

If Darwin's speculation is accurate, that all life forms have ultimately descended from, say, eight different progenitors, this means that there must have been at least eight different mergers of prokaryotes to form eukaryotes. This number is probably underestimated because it includes only those lines which survived. There may have been many more complete evolutionary paths that died off without a trace, leaving behind no fossils by which we could know of their existence.

Darwin's speculation of some eight ancestral lines is underestimated for another reason. If one is deliberately looking for a common ancestor for diverse species, there is a tendency to group animals together based on their similarities while overlooking significant differences which may suggest a different ancestral line. Similar body structures can be the result of similar evolutionary pressures rather than direct ancestry. Lemurs are a case in point. Based on superficial similarities, lemurs are believed to be the progenitors of monkeys and apes. Examination of the significant differences between these two groups of primates suggests otherwise.

Eyes

How eyes could have evolved in species is a question for which natural selection theory is unable to provide a satisfactory answer. Archeological evidence suggests that complex, image-forming eyes developed in over 50 species during a rapid burst of evolution in the Cambrian period (about 540 to 515 million years ago). These eyes are widely diverse in structure, each set being adapted to the unique requirements of the creatures which bear them. Does this mean that there was one ancestral line with over 50 branches, or does it mean that there could have been up to 50 different ancestral lines?

It makes sense that some primitive life forms had some surface cells that were more sensitive to light than others, and that successive generations of their offspring could develop improved structures for responding to light. However, to say that complex eye structures could have developed entirely on their own is a bit of stretch, especially considering that planning would have been required. If somehow an organism just developed eyes capable of sight (that its parents did not have), how would it see with them? If seeing eyes simply showed up at random or by the luck of a genetic lottery, how would the nervous system know that there were any images in the eyes or what to do with them? The only way that eyes could have evolved is if both they and the nervous system simultaneously and independently evolved with the same purpose in mind, rather like digging a tunnel from both sides of a mountain knowing that the two teams would meet precisely in the middle. This requires a level of planning of which individual organisms are incapable. The most viable explanation is that eyes were developed by purposefully switching on some inactive DNA modules that had been preprogrammed for the purpose, when the timing and situations were right for this to happen. Only a universal mind could have created the inactive future DNA potential, and only a universal mind could have turned it on when evolution required it – either through preprogramming or by direct intervention.

Lemurs

Lemurs belong to a primitive group of primates perhaps mistakenly called, "prosimian", because they are believed to be the ancestors of simian primates (apes and monkeys). Prosimian fossils predate simian fossils by about 20 million years; however, these two groups most probably have separate lineages.

Monkeys and lemurs have both similarities and major differences. Monkeys have prehensile tails capable of grasping; lemurs do not. Monkeys have dry noses surrounded by hair. Lemurs have moist, naked noses (like dogs) and longer snouts. Monkeys rely on sight for communication; lemurs rely heavily on smell for communication (and for choosing a mate). Monkeys have eyes like ours. Lemurs have reflective eyes that were designed for nocturnal living. Monkeys walk on two legs. Lemurs can stand up but always walk on all fours. Monkeys have a solid upper lip, while the lemur's upper lip is divided. Monkeys have larger brains and higher basal metabolic rates than lemurs.

Given the significant biological differences between these two kinds of primates, it is highly unlikely that the monkey group descended from the lemur group. This likelihood is made even smaller when one considers that there have been no fossilized remains found of any species in transition between the two distinctly different types. One wonders how many other cases of mistaken lineage have resulted from our inclination to group species together based on their similarities in an attempt to trace everything back to a common origin.

Platypus

The platypus is an unusual semi-aquatic egg laying mammal with four webbed feet and a large rubbery snout that superficially resembles a bird's beak. The male can deliver a venomous blow with its hind feet. Little is known about the ancestry of the platypus, except for fossilized remains of two similar animals that were found dating back 110 million years and 60 million years. It may be that the platypus has its own unique ancestral line unrelated to other mammals. Perhaps it is the only survivor of an independent evolutionary line resulting from a merging of two prokaryote cells to form a eukaryote cell.

Eels

The European (or true) eel differs from other fish in significant ways. It has a long ribbon-like body that pushes itself through the water in a wavelike manner, the way a snake moves on land. Adults live in fresh water for most of their lives, then every single one of them migrates to the Sargasso Sea (in the Western Atlantic between the Azores and the West Indies) to reproduce, a journey of up to 3,000 miles that can take up to seven months. The adults who have spawned remain in the Sargasso Sea, and their leaf shaped larvae are carried to the continental shelf of Europe by the Gulf Stream, a voyage that can take up to nine months.

To spawn, both salmon and eels return to the exact location where their individual lives began. The difference is that the lives of salmon begin in diverse locations in many parts of the world, but the lives of all eels begin in one very specific location.

The true eel has no ancestor. It did not evolve from any other known species. Two related questions: (1) how did the original eels get here, and (2) why does every eel undertake an arduous migration to a single remote location to reproduce? Every eel instinctively returns home, to its place of origin from which the Gulf Stream carried it away, to the single location where the lives of all eels began. Perhaps eels are a highly adaptable species whose DNA was brought here by scientists from another world – a living example of the directed panspermia in which Francis Crick, co-discoverer of DNA, believed.

Evolutionary theory cannot explain the origin of eels. The evidence suggests that DNA seeding is the most logical explanation for their origin, however far-fetched it may seem. The immortal words of the fictional detective, Sherlock Holmes, come to mind: *"How often have I told you that when you have eliminated the impossible, whatever remains, however improbable, must be the truth?"* If eels came to us by means of DNA seeding, then are there other species that have also arrived here by this or a similar method?

Jellies

Jellies (formerly called "jellyfish") are primordial marine animals having a gelatinous umbrella-shaped upper body with stinging tentacles, which they use passively like fishnets to trap their prey. These creatures first appeared in our seas about 650 million years ago, long before continents were formed. Jellies did not evolve from any other species.

Jellies are simple organisms with translucent bodies that are over 90 per cent water. They are unique in that they have no eyes, no ears, no brain, no heart, no blood, no spine, no vital organs, and no respiratory system. Jellies ingest their food chemically, which process starts by injecting their prey with venom. The only thing that jellies have in common with other animals is that they replicate by means of DNA – DNA that is unusual in that if it is injected into pig embryos, the piglets that are born glow in the dark when exposed to ultraviolet light.

Jellies have no ancestor and are unlike any other species on Earth. They must have come from somewhere else, but it is unlikely they could have survived space travel from another planet. At a certain stage in their development, jellyfish larvae transform into polyps that attach themselves to smooth rocks and lay dormant awaiting optimal conditions for their next stage of development. The only way that jellies could have arrived here from another planet is if that planet collided with Earth and showered it with marine rocks containing jellyfish polyps. Highly unlikely, because interplanetary collisions generate so much heat as to turn surfaces molten.

Another remote possibility is that two planets collided and one of them threw off from the opposite side of its impact some polyp-containing rocks that became meteorites, one or more of which eventually landed in Earth's primordial seas. Extremely unlikely that jellies could have survived such a voyage, which could have taken millions of years during which time the polyps would have been out of their natural environment of sea water.

When we eliminate the only two possibilities for how jellies could have arrived here by physical means, then we are left with the inescapable conclusion that they must have arrived here by nonphysical means. Jellies simply appeared in our oceans and could not have survived if there had not been a pre-existing food source for them, the most likely candidate being plankton.

Perhaps our primordial seas had become overrun with floating microscopic organisms and their waste products, and jellies were introduced as a cleanup operation to restore balance. If so, then this is an example of how a species can be created directly from nonphysical reality, bypassing the evolutionary process. If it happened once, it can happen again.

The physical and biological characteristics of jellies make them stand out like the proverbial sore thumb with respect to evolutionary pathways. What we do not know is if there have been other instances where nonphysical reality has created entirely new species with characteristics that blend in with those life forms that have evolved here. Perhaps we have been mistakenly attributing certain animals and plants to specific ancestral lines based on similarities, and evolutionary anomalies may come to light if we instead focus on unique differences.

Cyanobacteria

Cyanobacteria were the very first prokaryotic organisms on earth (a) to obtain their energy from photosynthesis, and (b) to give off gaseous oxygen as a byproduct of that photosynthesis. Their presence created what has been called the "great oxygenation event" that dramatically changed the composition of life on Earth by stimulating biodiversity and leading to the near extinction of anaerobic (oxygen intolerant) organisms.

Cyanobacteria created the conditions in the planet's early atmosphere that directed the evolution of aerobic metabolism and eukaryotic photosynthesis. Cyanobacteria changed the chemistry of the entire planet and ultimately made possible the evolution of all known plant and animal species.

The question is from what species did cyanobacteria evolve? How did oxygen-intolerant bacteria mutate into a species that gives off the oxygen that is poisonous to its parent species? It does not seem possible. A more plausible explanation is that cyanobacteria were transported here, perhaps by means of interplanetary collisions or meteorites.

Tardigrades

Tardigrades (water bears, moss piglets) are water-dwelling, segmented micro-animals about 0.07 mm. long, with eight legs, six of which are used for locomotion and the other two for grasping surfaces. Tardigrades can survive in extreme environments - withstanding temperatures from just above absolute zero (-270^0C) to well above the boiling point of water (100 °C), pressures about six times greater than those found in the deepest ocean trenches, ionizing radiation at doses hundreds of times higher than the lethal dose for a human, plus the vacuum of outer space. They can go without food or water for more than 10 years, drying out to the point where they are 3% or less water, only to rehydrate, forage, and reproduce.

Fossils of tardigrades have been found dating back to 530 million years ago. Unchanged in all that time, these hardy creatures have apparently survived five mass extinctions. The facts that they (a) have no known ancestors, (b) resemble no other creatures on Earth, and (c) can survive the extreme conditions of outer space suggest that they may have come to us from another planet, by clinging to the surfaces of meteorites. If so, then this species is both an exception to Earth's evolutionary patterns and evidence that life has also evolved elsewhere in the universe.

A Universe of Intention

Everything in the universe is intended.
Creation is the manifestation of intention.

If anything like a big bang happened, it could only have been intentional. If the universe is the result of continuous creation, this process could only have been set in motion intentionally. Whichever way it happened, the nonphysical must have been the cause. Thoughts create reality.

The orchestration of the physicality required to build a universe exceeds the capabilities of those physical components which participate in the process. The universe is beyond physics. Something nonphysical must be guiding or participating in the development of the universe.

Both theories, the big bang and continuous creation, lead us to the same ultimate cause – the nonphysical creates the physical. And it does so with precision. The universe is how it is because nonphysical reality intended it to be this way.

Why is there gravity? Because the nonphysical intended it so. If it were not for gravity, there would be no stars, no planets, no solar systems, no galaxies, and no life. Gravity is what has shaped the physical universe and holds it together, but gravity defies physical explanation. It simply is. Every bit of matter attracts every other bit of matter, just because it does. Gravity is a force of attraction that could only have been created by nonphysical reality.

This nonphysical reality must have done some very precise planning with respect to gravity. If gravity were too strong, the universe would have collapsed in on itself eons ago. If gravity were too weak, neither a big bang nor continuous creation could work – because all that could be produced by either event would be gases with no ability to clump together to form stars.

Why is there light? Because the nonphysical intended it so. There is no physical reason why electrons quantum leap between orbitals. But if they did not, no photons would be given off, the universe would be dark, there would be no photosynthesis and no atmospheric oxygen, and the only creatures that could have evolved would be microscopic organisms and perhaps some eyeless crustaceans.

Why are there different colors of light? Because the nonphysical intended it so. There is no physical reason why each element has a unique frequency at which its electrons quantum jump. But if they all jumped at the same frequency, everything we see would be the same color and we would have extreme difficulty in being able to distinguish one object from another, especially at a distance.

Electrons jump back and forth between orbitals – slipping momentarily into the nonphysical realm between jumps – and imparting to each kind of matter a unique vibration or frequency, depending on that substance's molecular structure. It is this energetic frequency of a material substance which gives it its uniqueness.

When an electron quantum jumps to a higher orbital, it absorbs energy in the form of photons. When it drops to a lower orbital, it gives off photons – which enable us to see the substance with our eyes and/or detect it with instrumentation. Both the absorbing and emitting of photons occur at the same unique frequency for each substance.

Why different frequencies? Because the nonphysical intended it so. Why do all electrons in every kind of matter not jump at the same frequency? If they did, then we would not be able to tell one substance from another. Everything would look and test the same. Our world would appear to be an unfathomable homogeneous blend of unknowns.

There is no physical reason why electrons jump orbitals. They simply do. There is no physical reason why electrons jump at different frequencies for different substances, yet that is exactly what they do. Every substance thus has a unique energy signature by which it can be identified. This is a deliberately orchestrated system of differentiation that is of significant benefit to us.

Violence Serves

We live in a violent universe. Asteroids, meteors, comets, planetary collisions, volcanic eruptions, and exploding stars pose continual threats to life. Ironically, life itself would not have been possible without all this planetary violence – which was part of the nonphysical master plan.

Stars and cosmic rays create elements by means of nuclear fusion and fission. Exploding stars eject element-containing gases and dust which become matter that accumulates to form planetary bodies – which, by crashing into each other, ensure that the 94 natural elements in our periodic table (20 of which are essential to life) are widely distributed throughout the universe. Icy comets bring water, dust particles, amino acids, sugars, and microorganisms. Microorganisms also hitch rides on meteorites and asteroids, thus enabling life forms to proliferate in new locations. Volcanic eruptions distribute minerals that enrich soils.

About 4.5 billion years ago, a planet about the size of Mars collided with newly forming Earth. The debris from this collision coalesced to become our moon. The gravitational pull of the moon stabilizes Earth's tilt, thus giving us habitable seasons. Without the moon, Earth would wobble so much on its axis as to induce frequent ice ages. In the early development of Earth, the gravitational pull of the moon also created tide pools whereby nutrients and microscopic life from the sea were distributed on nearby land. If our moon was created in this manner, then it is very likely that other moons have similarly been created in other solar systems and other galaxies.

About 2.3 billion years ago, cyanobacteria created the great oxygenation event that changed the chemistry of the entire planet and ultimately made possible the evolution of all known plant and animal species. These bacteria were most likely transported here by means of interplanetary collisions of some kind. This is the only logical explanation for how a species that was antagonistic to all existing species on the planet could have suddenly showed up and started producing oxygen by means of photosynthesis. Without cyanobacteria, Earth's atmosphere would be hostile to life as we know it.

The volcanic eruptions of 250 million years ago and the giant meteorite which crashed into earth 66 million years ago caused mass extinctions. In both cases, the death of species triggered bursts in evolution which created new species that would otherwise not have come to life. It was only after this first disaster that dinosaurs evolved. It was only after the second that advanced mammals, primates, and humans evolved.

Evidence suggests that there may have been five mass extinctions in Earth's history, on an average of every 100 million years. Each of these disasters wiped out most of the species on the planet – and in so doing, enabled the surviving species to take different evolutionary directions – thus creating cycles of life whereby the death of species becomes the birth of new species.

All this cosmological violence has helped to produce over 100 billion galaxies in the known universe, which collectively could be hosting as many as 50 sextillion habitable Earth-like planets – meaning planets that have stable orbits, stable tilt, water, oxygen, mineral-rich soils, amino acids, habitable temperatures, a magnetic field, and comparable gravity and atmospheric pressure. Galactic violence that the nonphysical deliberately set into motion thus serves the purpose of creating endless possibilities for life, including human-like life.

Bioprogramming

Why does the simple amoeba have 300 billion binary codes in its DNA and we humans have only three billion codes in ours? Because the nonphysical intended it so. This evidence suggests three things: (1) DNA is the bioequivalent of a computer program, (2) as beings evolve into higher life forms, they shed DNA that is no longer required, and (3) every living thing carries in its DNA genetic programming for future species to which it could evolve – an evolutionary reserve. These three factors are proof positive that DNA orchestration has been deliberately planned and could not possibly be the result of random events.

Enter Life

Chemistry cannot create biology. Primordial conditions can produce inert amino acids, but not life itself. Amino acids are dietary components that animals' bodies convert into protein tissues. It takes a digestive tract to do that, however. Amino acids cannot spontaneously become living organisms.

Whenever and wherever conditions and timing are suitable, nonphysical reality introduces life energy into the mix, along with the DNA program template upon which all life depends. Every time the nonphysical does this, new evolutionary pathways are created.

A Design

Nonphysical reality has set into motion physics, chemistry, and cosmology – the interplay of which has ultimately become the universe that we know. When and where conditions became receptive, the nonphysical also introduced life and evolution – from which have developed some eight million species of life on this planet alone. Gravity, planetary violence, light, life, DNA, evolution, photosynthesis – all were made manifest, either directly or indirectly, by the intentions of nonphysical reality.

We have no idea to what extent the nonphysical may also monitor the continuing development of its creations, or whether it simply observes their unfoldment. Perhaps nonphysical reality follows a "prime directive" as in one of the *Star Trek*-based television series which prohibits Starfleet personnel from interfering with the self determination of alien civilizations. Perhaps we (and others like us throughout the universe) are civilizations which the nonphysical leaves to our own devices and does not interfere.

An Infinite Universe

And the winner is ...

The universe is timeless, boundless, shapeless, limitless, infinite. The nonphysical deliberately made it this way by means of continuous creation. This is the only conclusion consistent with the evidence. Randomness had nothing to do with it. Neither did any alleged "big bang".

The notion that the universe could be a sphere created by a "big bang" of some 13.8 billion years ago defies logic in these ways:
- There is no point in time at which time began. The universe is timeless.
- There is no location at which the universe began. Without a universe, the concept of location makes no sense.
- The universe (everything which exists) has no boundary, spherical or otherwise. To have a boundary implies that something else (other than the universe) exists on the other side – a contradiction in terms.
- To have no boundary is to have no shape. To have no shape is to have no center.

If the universe had been caused by a big bang, it would be spherical in shape and have an identifiable center at the site of the explosion. Galaxies would be propelled away from this center, on radial paths. We would be able to know the position of the Milky Way galaxy in relation to the center of the universe – just as we know that Earth is about two-thirds of the way from the center of the Milky Way to its outer edge. If the cosmic microwave background had been caused by the big bang explosion, then we would see the CMB only when looking towards the universe's center. None of this is happening, however.

The universe has no shape and therefore no center. Without a central reference point, the concept of position has no meaning. We can tell only how far we are away from other galaxies. Neither they nor we are in any definable location.

> **Prediction**
> *By the year 2033, scientists will have discovered conclusive evidence of galaxies that pre-date the mythical big bang.*

is infinit space-time
consciousness no death

physical brain a receiver of non physical reality

physical brain in some space death

Consciousness

Everything which exists is the result of conscious thought.

The reality in which we live is consciousness based. Consciousness is an undeniable fact, a self-evident truth.

Without consciousness, I would not be writing, and you would not be reading my words. We would not be discussing these thoughts because thoughts themselves would not exist. Consciousness is a given.

We humans exist in and experience two realities simultaneously. Our bodies and brains exist in the physical reality. Our thoughts exist in a nonphysical reality. This is another self-evident truth. When you say, "This is my body," whomever you imagine yourself to be is speaking from the position of being other than your physical body, i.e. from a nonphysical state of being. When you say, "This is my brain," you are similarly speaking from a mental awareness that is other than your brain.

You are a conscious being who has a physical body. Through your five senses you experience physical reality. Through your thoughts you experience nonphysical reality. *to sense*

Mind and Brain

The mind is the power of consciousness, awareness, thought and volition. The mind exists only in nonphysical reality. The brain exists only in physical reality. There is no way that the experience of consciousness can be deduced from the physical properties of the brain.

The brain is a conduit for consciousness rather than the originator of that consciousness. Consciousness exists in the universe, irrespective of whether it is received or transmitted by a living brain.

The brain is the instrument that the nonphysical mind uses to direct the physical body. The brain does not create consciousness, rather it receives and processes thoughts received from consciousness. In this sense, the brain is like an energy transducer that converts nonphysical energy into biochemical electrical signals conveyed by enzymes and hormones.

The brain is a complex, sophisticated, and dynamic bio-processor, with the ability to change its own wiring connections. Neuroscience has been able to determine which lobes are involved in sensory information, memory, moods, emotions, sensing light, sensing smell and sound, processing complex stimuli, manipulating objects, and pain. The frontal lobe processes thoughts, but from where do these thoughts originate?

What science has been unable to find is any location in the brain that makes decisions. That is because decisions are made elsewhere. It is as if the brain is a bio-computer and your mind is the programmer. Your thoughts create molecules in the brain that communicate to the rest of your body.

There are numerous reports of people having near death experiences. What typically happens is that someone's body is clinically dead on an operating table for several minutes; and after being revived, the person describes phenomena experienced during the time in which the brain was incapable of any mental activity. It does not matter how similar or different these experiences are. It does not matter if these experiences are real or imaginary. If humans can have any kind of mental experience during an interval when their brains are documented to be nonfunctioning, this is proof positive that the mind is both (a) nonphysical and (b) exists independently of the brain. Consciousness exists in the absence of brain function.

We are not physical beings who happen to be conscious. We are conscious beings who happen to have physical bodies. Consciousness is not an attribute of being human. Being human is an attribute of consciousness. The implications for science are profound.

Consciousness and Evolution

As various species evolve into more advanced life forms, their brains necessarily evolve at the same pace. It cannot be otherwise. If brains did not evolve, evolution would have stopped with the primitive mammals of 200 million years ago, or perhaps with the worms of 540 million years ago. The more evolved a brain becomes, the more receptive and responsive it is to mental energy.

It is impossible for a physical brain to evolve into a nonphysical mind. The only plausible explanation for the mind-brain interface is that somehow the mind either selects or adapts the species that it wishes to occupy or direct.

Somewhere along the line a species evolved whose brain is a match for the human mind. This raises an interesting question: was there a child who can think that was born of parents who could not think? If so, what DNA did the child inherit that made this feat possible? Just as the brain cannot think, so also is there no "thinking" gene in our chromosomes.

All animals are aware of their environment. This is a requirement for survival. Animals mentally process whatever they see, hear or smell in terms of what is familiar and what is not. Whatever is unfamiliar receives their immediate full attention. At that point, the predator decides if this is an opportunity for food, and the prey decides if this is a danger from which it must flee. This decision is made instinctively and instantly, without deliberation (which could be fatal).

Domestic animals can recognize and respond to the names and commands we give them. Chimpanzees and the great apes can learn words, play with objects, play games, and communicate with us by means of sign language. Chimps can solve physical problems that are immediately present. They cannot, however, engage in abstract thought, theorize, or solve conceptual problems.

Animals have no choice but to live fully in whichever moment they find themselves. The kind of intelligence they have prevents them from regretting the past or worrying about the future, which kinds of deliberation could be fatal in life-or-death situations. Even the most intelligent of animals cannot mentally remove themselves from their present situations to deal with abstract concepts. In other words, animals cannot think. The consciousness which occupies an animal's body has a different experience than the consciousness which occupies a human body. Is it possible that consciousness chooses the kind of body it wishes to occupy or direct?

Simpler animal species can evolve into more aware, more complex species. As a brain becomes more sophisticated, the higher level of consciousness that it can receive from nonphysical reality. Consciousness is a primary. It is not a trait that can be passed on, neither from generation to generation, nor from species to species.

Consciousness and DNA

Wolves have in their untapped DNA the potential for their descendants to become every kind of dog imaginable – as demonstrated by over 400 breeds of *canis familiaris* that have been selectively bred from and can still mate with their ancestor, *canis lupis*. This feat is possible only if every species carries with it a broad range of DNA potential for future generations.

Identical twins have the same DNA, and sometimes these pairs choose different occupations requiring different skill sets. This suggests that one twin chose to pursue an interest that activated a portion of DNA that the other did not, and vice versa.

Each of us make conscious daily choices that tend to draw on some parts of our DNA and overlook others. Consider your parents, siblings, and children. If your family is typical, within those three generations there could be a wide range of abilities, interests, occupations, social skills, athletic prowess, and perhaps also intelligence. This may not be as much of a genetic lottery as it is individuals at some level choosing which parts of their DNA they wish to activate.

Universal Mind

Consciousness does not derive from brain function and is not part of the world in which the brain exists. Consciousness is part of the mind that infuses the universe.

Only a consciousness could have created the universe and orchestrated its continual development in such intricate and interrelated detail. Only a consciousness could have programmed DNA to provide for the evolution of endless possibilities of life forms – a consciousness that some philosophers have called a "universal mind." This universal mind permeates every aspect of existence.

Freeman Dyson spoke of a cosmic metaphysics of the mind in which the universe shows evidence of the operations of mind on three levels. First level is the process in quantum mechanics whereby matter is constantly making choices between alternate possibilities. Second level of mind is the level of direct human experience. Third level is that the universe itself may include a mental component or mental apparatus.

Of what does nonphysical reality consist? It has no physicality of any kind – no objects, no substances, no location, no boundaries, no time, and no motion. The nonphysical is the realm of pure creative energy, the energy of thought. The nonphysical is everywhere and nowhere.

The universal mind created both atoms and the entire cosmos and continually supplies them with nonphysical energy. The universal mind created life and continually supports it with life energy. Everything in the universe is thus a projection from thought.

Just as every invention began as a thought in someone's mind, the entire universe began as a thought in a universal mind. This universal mind decided to create the universe and planned how it would unfold in physical reality. The universe was thought into being.

I I is,
universal mind

Consciousness Creates

*Consciousness is the fundamental reality,
from which all other aspects of reality derive.*

Consciousness creates everything in our material world. Everything we see, hear, touch, taste, smell, detect, measure, observe, or experience (including ourselves) is either a manifestation of or a projection from consciousness. Thoughts become things. The universe has been thought into being.

Consciousness is the fundamental reality that cannot be explained by anything else. Universal consciousness creates the reality that we experience by means of our human consciousness. Both matter and mind are omnipresent in the universe.

All things are first created in nonphysical reality, then their counterparts are brought into physical existence. Nonphysical reality is the realm of pure consciousness that has no physical bounds and in which all thought/things are linked.

Thought is pure energy, without physicality. Thought is the energy of creation that transcends distance and time – and has also created distance and time.

Einstein once said, *"We are seeking for the simplest possible scheme of thought that will bind together the observed facts."* As it turns out, thought itself is the simplest possible scheme that explains the universe. Thought is the origin of everything.

A mathematical theory of everything is impossible because the creative force of consciousness cannot be quantified. It is possible, however, to express loosely the idea that everything is derived from consciousness in terms of mathematical symbols, such as the following:

$$\sum_{1}^{\infty} a \leq C$$

where a = any specific thing which exists, C = consciousness, ≤ means a subgroup of, and ∑ is a summation, in this case from one to infinity.

Interesting Parallels

1. Planets revolve in orbits around stars.
   Electrons revolve in orbitals around nuclei.

2. Over 99.999% of the universe is empty space.
   Over 99.999% of the atom is empty space.

3. Infinitely different arrangements of stars and galaxies.
   Infinitely different arrangements of atoms and molecules.

4. Cosmos is in constant motion.
   Constant motion within the atom.

5. Universe is continually being re-created.
   Atoms are continually being re-created.

6. Galaxies evolve.
   Species evolve.

7. Death of stars is birth of new stars.
   Death of species is birth of new species.

8. Cosmic web distributes energy from which galaxies are created.
   Meridians distribute energy from which the body is created.

9. A universal mind shapes reality.
   The human mind shapes our experience of that reality.

Glossary

Abiogenesis: the formation of living organism from non-living substances.

Adenine: *in biochemistry*, a purine derivative found in all living tissue as a component base of DNA or RNA (Symbol: **A**).

Amino acids: *in biochemistry*, any of a group of organic compounds containing both the carboxyl (COOH) group and the amino group (NH_2) group, occurring naturally in plant and animal tissues and forming the basic constituents of proteins.

Antimatter: *in physics*, hypothetical material composed of antiparticles which have the same mass as particles of ordinary matter but opposite electromagnetic properties.

Antiparticle: *in physics*, a hypothetical elementary particle having the same mass as a given particle but opposite electric or magnetic properties.

Argument: *in logic*, a set of (at least) two declarative sentences (or propositions) known as the premises, along with another declarative sentence which logically follows as the conclusion.

Ascetic: a person who practices severe self-discipline and abstains from all forms of pleasure, especially for religious or spiritual reasons.

Asteroid: any of the small planetary bodies found mainly between the orbits of Mars and Jupiter. Asteroids may release water as steam upon fiery impact with planets.

Astronomy: the study of the universe and its contents beyond the bounds of the earth's atmosphere.

Atom: the smallest particle of a chemical element that can take part in a chemical reaction; the basic unit of ordinary matter, made up of a tiny nucleus (consisting of protons and neutrons) surrounded by orbiting electrons.

ATP: adenosine triphosphate, a nucleotide important in living cells which, in breaking down to adenosine diphosphate, provides energy for physiological processes.

Atrophy: waste away from undernourishment, aging, or lack of use.

Attraction: *in physics*, the force by which bodies attract or approach each other.

Axiom: *in geometry*, a self-evident truth.

Bacteria: any of various groups of unicellular micro-organisms lacking organelles and an organized nucleus.

Being: (1) state of existence; anything that exists. (2) A human or other living entity.

Big bang: *in astronomy*, the violent explosion of all matter from a state of extremely high energy, postulated as the origin of the material universe.

Billion: a thousand million or 10^9.

Biochemistry: the study of the chemical and physiochemical processes of living organisms.

Biodiversity: variety of species.

Biology: the study of life and living organisms.

Black hole: a region of space having a gravitational field so intense that nothing, not even light, can escape from it.

Brain: an organ of soft nervous tissue contained in the skull of vertebrates, functioning as the coordinating center of sensation and nervous activity. The brain stores information and coordinates intellectual activity that originates in the mind. The brain executes mental decisions but does not make those decisions.

Chaos: *in physics*, the behavior of a system which is governed by cause and effect but is so unpredictable as to appear random.

Chemistry: the study of the elements, the compounds they form, and the reactions they undergo; the study of the composition, structure, properties, and changes of matter.

Ch'i: *in Chinese philosophy*, the nonphysical energy which composes all things, especially the vital life force residing within the breath and the body.

Chloroplast: *in biology*, a small organelle containing chlorophyll, found in plants undergoing photosynthesis.

Chromosome: *in biochemistry*, one of the threadlike structures, usually found in the cell nucleus, that carry genetic information in the form of genes.

Cloud: a visible mass of condensed watery vapor floating in the atmosphere high above the general level of the ground.

Codon: *in biochemistry*, a sequence of three nucleotides, forming a unit of genetic code in a DNA or RNA molecule.

Comet: a hazy celestial object with a nucleus of ice and dust surrounded by gas.

Complementarity: *in physics,* the concept that a single model may not be adequate to explain atomic systems in different experimental conditions.

Conclusion: a judgment reached by reasoning; *in logic*, a proposition that is reached by from given premises, the third and last part of a syllogism.

Consciousness: the state of awareness

Conservation of energy: a principle which states that energy can neither be created nor destroyed but only changed from one form to another, or transferred from one object to another.

Cosmology: the study of the origin and development of the universe.

Cosmos: the universe, as a complex and orderly, harmonious system.

Cyanobacteria: a type of bacteria that obtains its energy through photosynthesis.

Cytosine: *in biochemistry*, one of the principal component bases of the nucleotides and the nucleic acids, DNA and RNA, derived from pyrimidine (Symbol: C).

Cytoskeleton: *in biology*, a network of protein filaments and tubules giving shape and coherence to a living cell.

Deduce: draw a logical conclusion.

Define: give the exact meaning of (a word, etc.); specify, fix with precision.

Dichotomy: a division into two, especially a sharply defined one.

DNA: (*abbreviation*) deoxyribonucleic acid, the self-replicating material which is present in living organisms as a constituent of chromosomes and is the carrier of genetic information.

DNA polymerase: *in biochemistry*, any enzyme that catalyzes the formation of a DNA polymer.

Electricity: a form of energy resulting from the existence of charged particles (electrons, protons, etc.) either statically as an accumulation of charge or dynamically as a current.

Electromagnetic force: the force that arises between particles with electric charges.

Electromagnetism: the magnetic forces produced by electricity.

Electron: a stable subatomic particle with a charge of negative electricity, found in all atoms and acting as a primary carrier of electricity in solids.

Element: *in chemistry,* a substance that cannot be resolved by chemical means into simpler substances.

Elementary particle: a subatomic particle that is not decomposable into simpler particles.

Empirical: based or acted upon observation or experiment, not on theory; *in philosophy*, regarding sense data as valid information.

Energy: force, ability or capacity to produce an effect; *in physics*, the quantity of work a system is capable of doing.

Enigma: a puzzling, perplexing or unexplained thing.

Entanglement: a physical phenomenon that occurs when pairs or groups of particles are generated or interact in ways such that the quantum state each particle cannot be described independently – instead, a quantum state may be given for the system as a whole.

Entity: a thing with distinct and independent existence.

Eon: a very long or indefinite period of time.

Eukaryote: *in biology*, an organism consisting of a cell or cells in which the genetic material is DNA in the form of chromosomes contained within a distinct nucleus. The first such cell was believed to have been created by two simpler cells (prokaryotes) merging in such a way that one entered and became the nucleus of the other. Eukaryotes are the building block cells for all plants and animals.

Event: *in physics*, a single occurrence of a process, specified by its time and place. *In general*, a thing that happens or takes place.

Evolution: a process by which different kinds of organisms come into being by the differentiation and genetic mutation of earlier forms over successive generations.

Exist: to be real or actual; to have being.

Fact: a thing that is known to have occurred, to exist, or to be true.

Faraday cage: a grounded metal screen used for excluding electrostatic influences.

Force: effect exerted by one thing on another; *in physics*, an influence tending to cause the motion of a body.

Frequency: *in physics,* the rate of recurrence of a vibration or oscillation measured in cycles per second.

Geiger counter: a device for measuring radioactivity by detecting and counting ionizing particles.

Gene: a unit of heredity composed of DNA and forming part of a chromosome that determines a characteristic of an individual.

Gravity: *in physics*, a force of attraction between any particle of matter in the universe and any other.

Guanine: *in biochemistry*, a purine found in all living organisms as a component of DNA and RNA (Symbol: **G**).

Homeostasis: the tendency towards a relatively stable equilibrium between interdependent elements.

Humanoid: having human form or character.

Hypothesis: a proposition made as a basis for reasoning, without the assumption of its truth; a supposition made as a starting point for further investigation from known facts.

Hypothetical: supposed but not necessarily real or true.

Inductive reasoning: reasoning in which the premises seek to supply strong evidence for (but not absolute proof) of the truth of the conclusion. The conclusion of an inductive argument is either possible or probable, but not certain.

Infer: deduce or conclude from facts or reasoning.

Intention: a focused thought for creative purpose.

Interface: *in physics*, a surface forming a common boundary between two regions.

Jellyfish: a marine animal having a gelatinous umbrella-shaped upper body with stinging tentacles. Because this creature has nothing in common with fish, which are vertebrates with gills and fins. Scientists now prefer the term, "jellies".

Kinetic energy: energy which a body possesses by being in motion.

Law: a regularity in natural occurrences, especially as formulated or propounded in particular instances (*e.g., the laws of nature, the law of gravity*).

Life: the condition which distinguishes active animals and plants from inorganic matter, including the capacity for growth.

Light: natural electromagnetic radiation that stimulates sight and makes things visible. Light is given off by glowing objects due to thermal motion of atoms within them.

Light-year: *in astronomy*, the distance light travels in an average solar year, or approximately 5.88 trillion miles.

Logic: the science of reasoning, proof, thinking, or inference.

Manifest: to make apparent; to materialize, embody or incarnate.

Mass: *in physics*, the quantity of matter a body contains.

Material: formed or consisting of matter.

Mathematics: the abstract, deductive science of number, quantity, space, and arrangement studied in its own right or as applied to other disciplines, such as physics, engineering, etc.

Matter: that which has mass and occupies space. According to Einstein's famous $E = mc^2$ equation, energy can be transformed into matter, and vice versa. Matter is thus a slow moving, dense form of energy.

Mechanics: the branch of applied mathematics dealing with motion and tendencies to motion.

Meridian: any of the pathways in the body along which energy is said to flow, especially each of the 12 associated with specific organs for acupuncture.

Metaphysics: the branch of philosophy that deals with the first principles of things, including such concepts as being, knowing, substance, essence, cause, time and space.

Meteor: a small body of matter from outer space that becomes incandescent as a result of friction with the earth's atmosphere and is visible as a streak of light.

Meteorite: a rock or metal fragment from a meteor, of sufficient size to reach the earth's surface without burning up completely in the atmosphere.

Microcosm: a miniature representation.

Micron: one millionth of a meter.

Mind: the power of consciousness, awareness, thought and volition; the intellectual decision maker.

Mitochondrion: *in biology*, an organelle found in most eukaryotic cells, containing enzymes for respiration and energy production.

Momentum: *in physics*, the quantity of motion of a moving body, measured as a product of its mass and velocity.

Motion: the act or process of moving or changing position.

Mystic: a person who seeks by contemplation and self-surrender to obtain unity with or absorption into the ultimate reality.

Nanosecond: one billionth of a second.

Natural selection: the Darwinian theory of the survival and propagation of organisms best adapted to their environment.

Nebula: an interstellar cloud of dust, hydrogen, helium and other ionized gases that are the raw materials from which stars are born.

Neuroscience: the study of the structure and function of the brain and nervous system.

Neutron: *in physics*, an elementary particle of about the same mass as a proton but without an electric charge, present in all atomic nuclei except those of ordinary hydrogen.

Nonphysical: having no form or substance.

Nuclear fusion: a nuclear reaction in which nuclei of low atomic number fuse to form a heavier nucleus with the release of energy.

Nucleic acid: either of two complex substances (DNA or RNA) present in living cells, whose molecules consist of many nucleotides in a long chain.

Nucleoside: *in biochemistry*, an organic compound consisting of a purine or pyramidine base linked to a sugar (e.g., adenosine).

Nucleotide: *in biochemistry*, an organic compound that serves as a subunit of the nucleic acids, DNA and RNA.

Nucleus: *in physics*, the positively charged central core of an atom that contains most of its mass, consisting only of protons and neutrons, held together by the strong force; *in biology*, a large dense organelle in a eukaryotic cell, containing the genetic material.

Orbital: *in physics*, each of the actual or potential patterns of electron density which may be formed around an atomic nucleus by one or more electrons, represented as a wave function.

Organelle: *in biology*, any of various organized or specialized structures which form part of a cell.

Organic: *in chemistry,* (of a compound) containing carbon.

Organism: a living individual consisting of a single cell or of a group of interdependent parts sharing the life process.

Orgone energy: a universal life force conceived as a substratum in all of nature. Orgone is seen as a massless (pre-physical) substance more closely associated with living energy than with inert matter.

Origin: a beginning cause, starting point, or ultimate source.

Panspermia: the theory that life exists throughout the universe and is distributed by meteoroids, asteroids, comets, planetoids, and by spacecraft contaminated with microorganisms.

Paradox: a phenomenon that exhibits some contradiction or conflict with preconceived notions of what is reasonable or possible.

Particle: *in physics*, any of numerous subatomic constituents of the physical world that interact with each other, including electrons, neutrinos, photons and alpha particles. *In general*, a very small bit or piece of something, a tiny bit of matter.

Periodic table: an arrangement of chemical elements in order of increasing atomic number and in which elements of similar chemical properties appear at regular intervals.

Philosophy: the use of reason and argument in seeking truth and knowledge of reality, especially of the causes and nature of things and of the principles governing existence, the material universe, perception of physical phenomena and human behavior.

Photoelectric: marked by or using emissions of electrons from substances exposed to light.

Photon: a quantum of light or other electromagnetic radiation produced by the energizing of electrons. Photons have energy but no mass. Photons behave like particles and waves simultaneously. They have frequency, wavelength, and amplitude.

Photosynthesis: the process in which the energy of sunlight is used by organisms, especially green plants, to synthesize carbohydrates from carbon dioxide and water.

Physical: of or pertaining to matter, the world of the senses, or things material as opposed to things mental or spiritual.

Physics: the science dealing with the properties and interactions of matter and energy.

Physiology: the science that deals with the normal functioning of living organism and their parts.

Plankton: microscopic organisms drifting or floating in the sea or fresh water.

Plausible: *of an argument or statement*, seeming reasonable, believable, or probable.

Point: *in mathematics*, that which is conceived as having a position but no extent, magnitude, or dimension.

Polymer: a compound composed of one or more large molecules that are formed from repeated units of smaller molecules.

Postulate: assume, as a necessary condition, especially as a basis for reasoning.

Potentiality: the possibility of something developing or happening.

Premise: *in logic*, a previous statement or proposition from which another is inferred or follows as a conclusion; an assumption that something is true.

Primary: original, principal, foundational, underlying.

Primary energy: the omnipresent nonphysical energy that provides the underlying foundation for the universe. Primary energy has no cause and is the cause of everything. Being nonphysical, primary energy has no measurable characteristics.

Primate: any animal of the order, *Primates*, the highest order of mammals, including lemurs, apes, monkeys and humans.

Primordial: existing at or from the beginning; original, fundamental.

Probability: the likelihood of something happening.

Progenitor: the ancestor of a person, animal, or plant.

Prokaryote: *in* biology, a single celled organism which has neither a distinct nucleus with a membrane nor other specialized organelles. Bacteria and blue green algae are examples of prokaryotes.

Prosimian: a primitive primate believed to be the ancestors of apes and monkeys.

Protein: any of a group of organic compounds composed of one or more chains of amino acids and forming an essential part of all living organisms.

Proton: *in physics*, a stable elementary particle with a positive electrical charge, equal to that of an electron, and accounting for roughly half the particles in the nuclei of most atoms.

Purine: any of a group of compounds with a similar structure, such as the adenine and guanine nucleotides, which form uric acid on oxygenation.

Pyrimidine: any of a group of cyclic organic compounds with a similar structure, including the nucleotide constituents, uracil, thymine, and cytosine.

Quantum: a discrete quantity or packet of energy proportional in magnitude to the frequency of radiation it represents.

Quantum jump: *in physics*, an abrupt transition in an atom or molecule from one quantum state to another.

Quantum mechanics: *in physics*, a mathematical theory dealing with the motion and interaction of subatomic particles and incorporating the concept that these particles can also be waves.

Quasar: a compact region surrounding a black hole that gives off intense radiation.

Radiation: *in physics*, the emission of energy as electromagnetic waves or as moving particles.

Radioactivity: the spontaneous disintegration of atomic nuclei, with the mission of penetrating radiation or particles.

Reasoning: forming or reaching conclusions by connected thought.

Redshift: *in physics*, the phenomenon by which light or other electromagnetic radiation from an object is increased in wavelength (with a corresponding decrease in frequency). This term was named for visible light that is being shifted to the red end of the spectrum; however, it applies to all forms of radiation.

Reductio ad absurdum: a method of proving the falsity of a premise by showing that its logical consequence is absurd.

Relativity: *in physics*, a theory based on the principle that all motion is relative and that light has a constant velocity; **general theory of relativity**: a theory extending this to gravitation and accelerated motion.

RNA: (*abbreviation*) ribonucleic acid, a nucleic acid present in all living cells, especially in ribosomes where it is involved in protein synthesis.

Science: the systematic study of the structure and behavior of the physical and natural world through observation and experiment.

Scientific method: systematic observation, measurement and experiment, and the formulation, testing, and modification of hypotheses.

Sound wave: a longitudinal pressure wave in an elastic medium, such as air, that propagates audible sound.

Source: a beginning, cause, or reason from which something originates.

Space: boundless three-dimensional extent in which objects have relative position and direction. Important concept to understanding the physical universe.

Spectrophotometer: an instrument for measuring the intensity of light in various parts of the spectrum, especially as transmitted or emitted by a substance or a solution at a particular wavelength.

Stem cell: *in biology*, an undifferentiated cell from which specialized cells developed.

Superposition state: *in quantum mechanics*, the condition in which a subatomic particle exists in all possible states simultaneously.

Template: a pattern or gauge; *in biochemistry*, a nucleic acid molecule that acts as a pattern for the sequence of assembly of a protein, nucleic acid, or other large molecule.

Theorem: a general proposition not self-evident but proved by a chain of reasoning; a truth established by means of accepted truths.

Theory: a supposition or system of ideas explaining something, especially one based on general principles independent of the things to be explained.

Think: to create and develop abstract concepts; to deduce or infer.

Thought: product of mental activity, such as a concept, idea, or image.

Thought experiment: a mental assessment of the implications of a hypothesis.

Thymine: *in biochemistry*: a pyrimidine found in all living tissue as a component base of DNA (Symbol: **T**).

Time: the continuous duration of existence seen as a series of events progressing from the past through the present into the future.

Transduce: to convert energy into a different medium or form of energy.

Transposon: in biology, a mobile segment of DNA that can replicate and insert copies of DNA at random or pre-selected sites in the same or a different chromosome.

Uncertainty principle: the more precisely the position of some particle is determined, the less precisely its momentum can be measured, and vice versa.

Universal: affecting everything.

Universe: all existing things.

Vesicle: *in biology*, a small fluid-filled bladder, sac or vacuole.

Vibration: moving continuously and rapidly to and fro; oscillating, reciprocating periodic motion of a body forced from its position of equilibrium.

Virtual: that is such in essence or effect, although not recognized as such in name or according to strict definition.

Virtual particle: *in quantum mechanics*, a hypothetical particle that occurs over a very short interval and can never be directly detected, but whose supposed existence has measurable effects.

Void: an empty space or vacuum.

Weight: *in physics*, the force experienced by a body as a result of the earth's gravitational pull; the force exerted by a gravitational field. It is proportional to, but not the same as its mass.

About the Author

David Rowland is self-educated in several fields, a member of the Royal Astronomical Society of Canada, and a philosopher in the original sense of the word, meaning "lover of wisdom". David is also Canada's foremost expert in holistic nutrition, having written 10 books on this subject, some of which are used as texts in nutrition schools.

Other Books by David Rowland

WHAT WE KNOW ABOUT THE UNIVERSE THAT ISN'T SO
Science is at a Crossroads

QUANTUM THEORY DEMYSTIFIED
Continuous Creation of the Atom

NUTRITIONAL SOLUTIONS FOR 88 CONDITIONS
Correct the Causes

BYPASS THE BYPASS
Restore Circulation Without Surgery

Made in the USA
Coppell, TX
22 December 2019

13668206R00105